Visual Object Recognition

Synthesis Lectures on Artificial Intelligence and Machine Learning

Editors
Ronald J. Brachman, *Yahoo Research*
Thomas Dietterich, *Oregon State University*

Representation Discovery using Harmonic Analysis
Sridhar Mahadevan
2008

Essentials of Game Theory: A Concise Multidisciplinary Introduction
Kevin Leyton-Brown and Yoav Shoham
2008

A Concise Introduction to Multiagent Systems and Distributed Artificial Intelligence
Nikos Vlassis
2007

Intelligent Autonomous Robotics: A Robot Soccer Case Study
Peter Stone
2007

Visual Object Recognition
Kristen Grauman and Bastian Leibe

ISBN: 978-3-031-00425-4   paperback
ISBN: 978-3-031-01553-3   ebook

DOI 10.1007/978-3-031-01553-3

A Publication in the Springer series
SYNTHESIS LECTURES ON ARTIFICIAL INTELLIGENCE AND MACHINE LEARNING

Lecture #11
Series Editors: Ronald J. Brachman, *Yahoo Research*
                Thomas Dietterich, *Oregon State University*
Series ISSN
Synthesis Lectures on Artificial Intelligence and Machine Learning
Print 1939-4608   Electronic 1939-4616

Visual Object Recognition

Kristen Grauman
University of Texas at Austin

Bastian Leibe
RWTH Aachen University

SYNTHESIS LECTURES ON ARTIFICIAL INTELLIGENCE AND MACHINE LEARNING #11

ABSTRACT

The visual recognition problem is central to computer vision research. From robotics to information retrieval, many desired applications demand the ability to identify and localize categories, places, and objects. This tutorial overviews computer vision algorithms for visual object recognition and image classification. We introduce primary representations and learning approaches, with an emphasis on recent advances in the field. The target audience consists of researchers or students working in AI, robotics, or vision who would like to understand what methods and representations are available for these problems. This lecture summarizes what is and isn't possible to do reliably today, and overviews key concepts that could be employed in systems requiring visual categorization.

KEYWORDS

global representations versus local descriptors; detection and description of local invariant features; efficient algorithms for matching local features, tree-based and hashing-based search algorithms; visual vocabularies and bags-of-words; methods to verify geometric consistency according to parameterized geometric transformations; dealing with outliers in correspondences, RANSAC and the Generalized Hough transform; window-based descriptors, histograms of oriented gradients and rectangular features; part-based models, star graph models and fully connected constellations; pyramid match kernels; detection via sliding windows; Hough voting; Generalized distance transform; the Implicit Shape Model; the Deformable Part-based Model

Contents

Preface

This lecture summarizes the material in a tutorial we gave at AAAI 2008 [Grauman and Leibe, 2008][1]. Our goal is to overview the types of methods that figure most prominently in object recognition research today, in order to give a survey of the concepts, algorithms, and representations that one might use to build a visual recognition system. Recognition is a rather broad and quickly moving field, and so we limit our scope to methods that are already used fairly frequently in the literature. As needed, we point the reader to outside references for more details on certain topics. We assume that the reader has basic familiarity with machine learning algorithms for supervised classification and some background in low-level image processing.

Outline: The text is divided primarily according to the two forms of recognition: specific and generic. In the first chapter, we motivate those two problems and point out some of their main challenges. Chapter 2 then overviews the problem of specific object recognition, and the global and local representations used by current approaches. Then, in Chapters 3 through 5, we explain in detail how today's state-of-the-art methods utilize local features to identify an object in a new image. This entails efficient algorithms for local feature extraction, for retrieving candidate matches, and for performing geometric verification. We wrap up our coverage of specific objects by outlining example end-to-end systems from recent work in Chapter 6, pulling together the key steps from the prior chapters on local features and matching.

Chapter 7 introduces the generic (category-level) recognition problem, and it is followed by a discussion of the window-based and part-based representations commonly used to describe object classes in Chapter 8. Having overviewed these models, we then describe how they are used to localize instances of a given category in new images in Chapter 9. Then in Chapter 10, we describe how the window-based or part-based models are typically learned, including both the standard training data preparation process as well as the applicable classifiers. As with the specific object recognition section, we conclude this portion of the lecture with a chapter outlining some state-of-the-art end-to-end systems that utilize the basic tools we have introduced (Chapter 11).

Finally, we end the lecture with a chapter overviewing some important considerations not discussed in detail in the above (such as context models, or concurrent segmentation and recognition),

---

[1]The authors' website associated with that tutorial provides pointers to online resources relevant to the text, including datasets, software, and papers.

and provide pointers on these and other interesting subproblems that are being actively studied in the recognition community.

Kristen Grauman and Bastian Leibe
March 2011

Acknowledgments

This project began after we gave a tutorial on the subject at AAAI 2008; we thank the conference organizers for hosting that event and the attendees for their questions and feedback. We also would like to thank Derek Hoeim and Mark Everingham for their helpful comments that improved the organization of this manuscript. We also thank the many colleagues who kindly gave permission to use their figures in this book (as listed in the figure credits).

Thanks to the following Flickr users for sharing their photos under the Creative Commons license: belgianchocolate, c.j.b., edwinn.11, piston9, staminaplus100, rickyrhodes, Rick Smit, Krikit, Vanessa Pike-Russell, Will Ellis, Yvonne in Willowick Ohio, robertpaulyoung, lin padgham, tkcrash123, jennifrog, Zemzina, Irene2005, CmdrGravy, soggydan, cygnus921, Around the World Journey 2009-20xx - mhoey, jdeeringdavis, yummiec00kies, kangotraveler, laffy4k, jimmyharris, Bitman, http2007, aslakr, uzvards, edg1, chris runoff, Jorge-11, Arenamontanus, and bucklava.

Kristen Grauman's work on this project was supported in part by NSF CAREER IIS-#0747356, a Microsoft Research New Faculty Fellowship, and the Henry Luce Foundation. Bastian Leibe's work on the project was supported in part by the UMIC cluster of excellence (DFG EXC 89).

Kristen Grauman and Bastian Leibe
March 2011

Figure Credits

Figure 1.4   Figures are from the MNIST, COIL, ETH-80, and PASCAL VOC 2008 data sets, some of which are based on Flickr photos. Others are based on Besl and Jain [1985]; Grimson et al. [1991]; Fan et al. [1989]; Viola and Jones [2001]; and Lowe [2004].

Figure 2.2   Courtesy of David Lowe.

Figure 3.1   Courtesy of Krystian Mikolajczyk.

Figure 3.2   Courtesy of Denis Simakov.

Figure 3.3   Courtesy of Krystian Mikolajczyk.

Figure 3.5   Courtesy of Krystian Mikolajczyk.

Figure 3.6   From Tuytelaars and Mikolajczyk [2007]. Used with permission.

Figure 3.7   Courtesy of Krystian Mikolajczyk, and from Tuytelaars and Mikolajczyk [2007].

Figure 3.8   Based on Lowe [1999] and Lowe [2004].

Figure 3.9   Based on Bay et al. [2008].

Figure 4.2   Courtesy of The Auton Lab.

Figure 4.3   From Kulis et al. [2009]. Copyright © 2009 IEEE. Used with permission.

Figure 4.5   From Sivic and Zisserman [2003]. Copyright © 2003 IEEE. Copyright © 2003 IEEE. Used with permission.

Figure 4.7   Courtesy of Ondrej Chum.

Figure 5.1   Courtesy of Krystian Mikolajczyk.

Figure 5.2   From Lowe [1999]. Copyright © 1999 IEEE. Used with permission.

Figure 5.3   Courtesy of Jinxiang Chai.

Figure 5.4   Courtesy of Svetlana Lazebnik, David Lowe, and Lowe [2004], left, and from Leibe, Schindler and Van Gool [2008], right, Copyright © 2008 Springer-Verlag. Used with the kind permission of Springer Science+Business Media.

Figure 6.1   From Tuytelaars and Van Gool [2004], Copyright © 2004 Springer-Verlag. Used with the kind permission of Springer Science+Business Media.

Figure 6.2   Courtesy of Matthew Brown and from Brown and Lowe [2003, 2007]), Copyright © 2007 Springer-Verlag. Used with the kind permission of Springer Science+Business Media.

Figure 6.3   Based on Lowe [1999].

Figure 6.4   From Philbin et al. [2007], Copyright © 2007 IEEE. Used with permission.

Figure 6.5   From Gammeter et al. [2009], Copyright © 2009 IEEE. Used with permission.

Figure 8.1   Based on Turk and Pentland [1992], Torralba [2003], and Bosch et al. [2007b], and courtesy of Svetlana Lazebnik.

Figure 8.2   Based on Lowe [1999], Bay et al. [2006], and Viola and Jones [2004], and courtesy of Svetlana Lazebnik.

Figure 8.4   Courtesy of Rob Fergus and Courtesy of The CalTech256 Dataset.

Figure 8.5   Based on Carneiro and Lowe [2006].

Figure 8.6   From Fergus et al. [2003]. Copyright © 2003 IEEE. Used with permission.

Figure 8.7   From Fergus et al. [2003]. Copyright © 2003 IEEE. Used with permission.

Figure 9.3a   Left image from Viola and Jones [2001]. Copyright © 2001 IEEE. Used with permission.

Figure 9.3b   Courtesy of Derek Hoiem.

Figure 10.1a   From Yuille et al. [1992].

Figure 10.1b   From Cootes et al. [2001]. Copyright © 2001 IEEE. Used with permission.

Figure 10.1c   From Viola and Jones [2001]. Copyright © 2001 IEEE. Used with permission.

Figure 10.1d   From Vijayanarasimhan and Grauman [2011]. Copyright © 2011 Springer-Verlag. Used with the kind permission of Springer Science+Business Media.

Figure 10.1e   From the PASCAL VOC data set.

Figure 10.1f   Courtesy of the PASCAL VOC image dataset and The Caltech256 dataset, respectively.

Figure 10.2   Courtesy of Alex Berg, courtesy of Serge Belongie, and courtesy of Christian Wallraven. Also, from Grauman and Darrell [2006a]. Copyright © 2006 IEEE. Used with permission. And from Ling and Jacobs [2007a]. Copyright © 2007 IEEE. Used with permission.

Figure 10.4   From Lazebnik et al. [2006]. Copyright © 2006 IEEE. Used with permission.

Figure 10.5a   From Berg et al. [2005]. Copyright © 2005 IEEE. Used with permission.

Figure 10.5b   Courtesy of Alex Berg.

Figure 10.6   Left is from Quack et al. [2007], Copyright © 2006 Association for Computing Machinery. Top right is from Sivic and Zisserman [2004], Copyright © 2004, IEEE. Bottom right is from Lazebnik et al. [2004], Copyright © 2004, British Machine Vision Association.

Figure 10.7   From Leibe, Leonardis and Schiele [2008]. Copyright © 2008 Springer-Verlag. Used with the kind permission of Springer Science+Business Media.

Figure 10.8   From Felzenszwalb, Girshick, McAllester and Ramanan [2010]. Copyright © 2010 IEEE. Used with permission.

Figure 11.4   Courtesy of Navneet Dalal.

Figure 11.6   From Leibe, Leonardis and Schiele [2008]. Copyright © 2008 Springer-Verlag. Used with the kind permission of Springer Science+Business Media.

Figure 11.7   From Leibe, Leonardis and Schiele [2008]. Copyright © 2008 IEEE. Used with the kind permission of Springer Science+Business Media.

Figure 11.8   Based on Thomas et al. [2009a].

Figure 11.9   From Leibe et al. [2005]. Copyright © 2005 IEEE. Used with permission.

Figure 11.10   From Felzenszwalb, Girshick, McAllester and Ramanan [2010]. Copyright © 2010 IEEE. Used with permission.

Figure 11.11   From Felzenszwalb, Girshick, McAllester and Ramanan [2010]. Copyright © 2010 IEEE. Used with permission.

Figure 12.1a   Courtesy of The CalTech256 Dataset.

Figure 12.1b   From the PASCAL VOC data set.

Figure 12.1c   Courtesy of The LabelMe Dataset.

Figure 12.1d   From the MSRC dataset.

CHAPTER 1

Introduction

1.1 OVERVIEW

Recognition is the core problem of learning visual categories and then identifying new instances of those categories. Most any vision task fundamentally relies on the ability to recognize objects, scenes, and categories. Visual recognition itself has a variety of potential applications that touch many areas of artificial intelligence and information retrieval—including, for example, content-based image search, video data mining, or object identification for mobile robots.

Recognition is generally considered by vision researchers as having two types: the specific case and the generic category case (see Figure 1.1). In the specific case, we seek to identify instances of a particular object, place, or person—for example, Carl Gauss's face, the Eiffel Tower, or a certain magazine cover. In contrast, at the category level, we seek to recognize different instances of a generic category as belonging to the same conceptual class—for example, buildings, coffee mugs, or cars. In either case, we can expect that there may be variations in appearance among different instances of the same class or object. This lecture explains the basic algorithms for both forms of recognition, based on advances in the computer vision literature in the last eight to ten years.

When trying to address the recognition task, an important first question to ask is what sorts of categories can be recognized on a visual basis? In order to answer this question, it is useful to look at how humans organize knowledge at different levels. This question has received considerable attention in Cognitive Psychology [Brown, 1958]. Taking an example from Brown's work, a dog can not only be thought of as a *dog*, but also as a *boxer*, a *quadruped*, or in general an *animate being* [Brown, 1958]. Yet, dog is the term (or level in the semantic hierarchy) that comes to mind most easily, which is by no means accidental. Experiments show that there is a basic level in human categorization at which most knowledge is organized [Rosch et al., 1976]. According to Rosch et al. [1976] and Lakoff [1987], this basic level is also

- the highest level at which category members have similar perceived shape.

- the highest level at which a single mental image can reflect the entire category.

- the highest level at which a person uses similar motor actions for interacting with category members.

- the level at which human subjects are usually fastest at identifying category members.

- the first level named and understood by children.

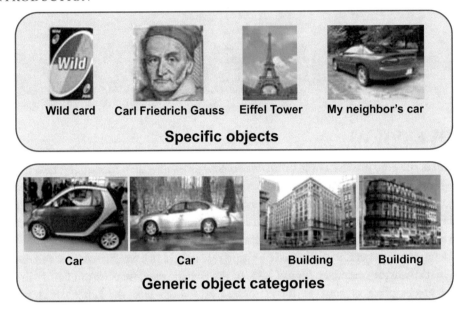

Figure 1.1: The recognition problem includes identifying instances of a particular object or scene (top) as well as generalizing to categorize instances at a basic level (bottom). This lecture overviews methods for specific object recognition in Chapters 2 through 6, followed by generic object categorization and detection in Chapters 7 through 11.

Together, those properties indicate that basic-level categories are a good starting point for visual classification, as they seem to require the simplest visual category representations. Category concepts below this basic level carry some element of specialization (*e.g.*, different sub-species of dogs or different models of cars) down to an individual level of specific objects (*e.g.*, *my dog* or *my car*), which require different representations for recognition. Concepts above the basic level (*e.g.*, *quadruped* or *vehicle*) make some kind of abstraction and therefore typically require additional world knowledge on top of the visual information.

In this lecture, we first overview algorithms for recognizing specific objects, and then describe techniques for recognizing basic-level visual categories. We explicitly do not model functional categories (*e.g.*, "things you can sit on") and ad-hoc categories (*e.g.*, "things you can find in an office environment") [Barsalou, 1983]. Though those categories are also important, they exist only on a higher level of abstraction and require a more advanced degree of world knowledge and experience living in the real world.

In computer vision, the current standard pipeline for specific object recognition relies on a matching and geometric verification paradigm. In contrast, for generic object categorization, it often also includes a statistical model of appearance or shape learned from examples. For the categorization problem, learning visual objects entails gathering training images of the given category, and then

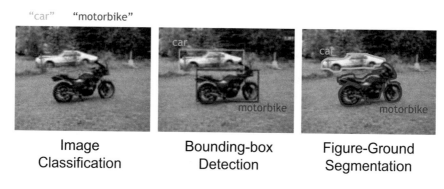

Image Classification   Bounding-box Detection   Figure-Ground Segmentation

Figure 1.2: The task of a given recognition system can vary in the level of detail, ranging from an image-level label (left), to a bounding box detection (center), to a complete pixel-level segmentation of the object (right).

extracting or learning a model that can make new predictions for object presence or localization in novel images. Models are often constructed via supervised classification methods, with some specialization to the visual representation when necessary.

The type of training data required as well as the target output can vary depending on the detail of recognition that is required. Specifically, the target task may be to *name* or categorize objects present in the image, to further *detect* them with coarse spatial localization, or to *segment* them by estimating a pixel-level map of the named foreground objects and the background (see Figure 1.2).

1.2   CHALLENGES

Matching and learning visual objects is challenging on a number of fronts. Instances of the same object category can generate very different images, depending on confounding variables such as illumination conditions, object pose, camera viewpoint, partial occlusions, and unrelated background "clutter" (see Figure 1.3). Different instances of objects from the same category can also exhibit significant variations in appearance. Furthermore, in many cases appearance alone is ambiguous when considered in isolation, making it necessary to model not just the object class itself, but also its relationship to the scene context and priors on usual occurrences.

Aside from these issues relating to robustness, today's recognition algorithms also face notable challenges in computational complexity and scalability. The fact that about half of the cerebral cortex in primates is devoted to processing visual information gives some indication of the computational load one can expect to invest for this complex task [Felleman and van Essen, 1991]. Highly efficient algorithms are necessary to accommodate rich high-dimensional image representations, to search large image databases, or to extend recognition to thousands of category types. In addition, scalability concerns also arise when designing a recognition system's training data: while unambiguously labeled

Figure 1.3: Images containing instances of the same object category can appear dramatically different; recognition methods must therefore be robust to a variety of challenging nuisance parameters.

image examples tend to be most informative, they are also most expensive to obtain. Thus, methods today must consider the trade-offs between the extent of costly manual supervision an algorithm requires versus the advantages given to the learning process.

1.3   THE STATE OF THE ART

In spite of these clear challenges, object recognition research has made notable strides within the last decades. Recent advances have shown the feasibility of learning accurate models for some well-defined object categories—from early successes for detecting a particular class of interest (notably faces and pedestrians), to indexing methods for retrieving instances of a particular object, on to models handling tens of generic object categories. Within the last few years, thanks in part to work developing standardized benchmark databases and recognition challenges, researchers have set their sights on more complex multi-class problems that involve detecting object classes within realistic cluttered backgrounds, or categorizing objects from hundreds to thousands of categories. Figure 1.4 illustrates the evolution of the field's attention by showing snapshots of the kinds of datasets that have been employed over the years, ranging from the late 1980's until today.

Figure 1.4: A sampling of the kind of image data recognition methods have tackled over the years, approximately ordered from top left to bottom right. As this imagery reflects, the task has evolved from single object instances and prepared data for particular classes, to include fairly natural images with generic object categories and clear viewpoint changes. Figures are from the MNIST, COIL, ETH-80, and PASCAL VOC 2008 data sets, some of which are based on Flickr photos. Others are based on Besl and Jain [1985]; Grimson et al. [1991]; Fan et al. [1989]; Viola and Jones [2001]; Philbin et al. [2007]; and Lowe [2004].

While a variety of approaches have been explored, they all must make a few common choices: How will the images and models be represented? Using that representation, how is a category learned? Given a novel image, how is categorization or detection carried out? In this tutorial, we overview algorithms for visual recognition and focus on the possible answers to these questions in the light of recent progress in the field.

CHAPTER 2

Overview: Recognition of Specific Objects

The first half of this book will address the problem of recognizing *specific objects*. For this, we will give an overview of suitable representations and the corresponding recognition pipelines.

The second half of the book, starting from Chapter 7, will then introduce methods for recognizing *generic object categories*.

2.1 GLOBAL IMAGE REPRESENTATIONS

Perhaps the most direct representation of an appearance pattern is to write down the intensity or color at each pixel, in some defined order relative to a corner of the image (see Figure 2.1 (a)). If we can assume that the images are cropped to the object of interest and rather aligned in terms of pose, then the pixel reading at the same position in each image is likely to be similar for same-class examples. Thus, the list of intensities can be considered a point in a high-dimensional appearance space where the Euclidean distances between images reflect overall appearance similarity.

In what is among the earliest work proposing a statistical model for object appearance, Turk and Pentland considered a related vector space model for images [Turk and Pentland, 1992]. Their *Eigenface* approach uses Principal Components Analysis (PCA) to recover the principal directions of variation in the space of all face images, thereby reducing the high-dimensional pixel lists to a much more compact encoding of the key appearance attributes. To recognize a novel image window, it is projected onto the subspace spanned by the eigenfaces, and then mapped to the face label for the training instance that is nearest to it in the lower-dimensional subspace.

Murase and Nayar [1995] generalized this idea to arbitrary 3D objects and constructed a system that could recognize and distinguish 100 objects in real-time [Nayar et al., 1996]. For varying poses (rotations around the object's vertical axis), the projected images lie on a manifold in the eigenspace. Murase & Nayar's system achieves recognition invariant to object pose by projecting the (pre-segmented) test image onto the closest point on the manifold. By interpolating the manifold in-between training images, a highly accurate pose estimation can be achieved.

Belhumeur and Kriegman [1996] extend the recognition approach into a different direction. The subspace created by PCA is designed to minimize the reconstruction error by projecting images onto their dimensions of maximum variability. While this projection creates a subspace of reduced dimensionality with minimal loss of image information, the corresponding projection does not consider object class information and may therefore result in suboptimal discrimination between

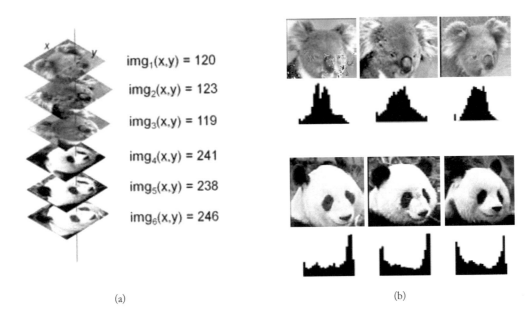

(a)   (b)

Figure 2.1: A holistic description of a window's appearance can be formed by its ordered (a) or unordered (b) set of intensities. In (a), the images' ordered lists of intensities are considered as high-dimensional points; very similar images will have close values in each dimension. In (b), the images are mapped to grayscale histograms; note how the distribution of intensities has a distinct pattern depending on the object type.

different model objects. Belhumeur and Kriegman [1996] therefore propose to optimize class separability instead by working in a subspace obtained by Fisher's Linear Discriminant Analysis (often called FLD or LDA, depending on the community). The resulting *Fisherfaces* approach achieves better discrimination capabilities than PCA-based methods and can be applied to construct subspaces optimized for specific discrimination tasks (such as distinguishing people wearing glasses from people wearing no glasses).

A limitation of ordered global representations is their requirement that the 2D patterns be closely aligned; this can be expensive to ensure when creating a labeled training set of images. For example, if a face class is represented by ordered lists of intensities taken from examples of frontal faces, with even a slight shift of an example, the nose pixels will no longer be in the "right" position, and the global representation will fail.

Given an image window, we can alternatively form a simple holistic description of its pixels using the distribution of its colors (or intensities, for grayscale images; see Figure 2.1(b)). A color histogram measures the frequency with which each color appears among all the window's pixels, where discrete color types are defined as bins in the color space, *e.g.*, RGB or Lab. Being orderless, the histogram offers invariance to viewing conditions and some tolerance for partial occlusions. This insensitivity makes it possible to use a small number of views to represent an object, assuming a closed-world pool with distinctly colored objects. The first approach making use of color histograms for specific object recognition was proposed by Swain and Ballard [1991]. The idea was subsequently extended by Schiele and Crowley [2000] to multidimensional histograms of local image neighborhood operators (ranging from simple derivatives to more complex derivative combinations) extracted at multiple scales. Those histograms effectively capture the probability distribution of certain feature combinations occurring on the object. In addition to a global recognition approach by matching entire histograms, Schiele and Crowley [2000] also propose a local method that estimates the maximum a-posteriori probability of different model objects from a few sampled neighborhoods in the test image.[1]

Most of the global representations described above lead to recognition approaches based on comparisons of entire images or entire image windows. Such approaches are well-suited for learning global object structure, but they cannot cope well with partial occlusion, strong viewpoint changes, or with deformable objects. In the following, we describe an alternative representation based on local features that has in the meantime become the standard for specific object recognition tasks.

2.2   LOCAL FEATURE REPRESENTATIONS

The successful development of local invariant feature detectors and descriptors has had a tremendous impact on research in object recognition. Those features have made it possible to develop robust and efficient recognition approaches that can operate under a wide variety of viewing conditions and under partial occlusion.

In the following chapters, we give an overview of the basic processing pipeline underlying many of those approaches for the case of specific object recognition. A visualization of the basic recognition procedure is shown in Figure 2.2. Given a model view of a (rigid) object, the task is to recognize whether this particular object is present in the test image and, if yes, where it is precisely located and how it is oriented. This task is addressed by representing the image content by a collection of local features that can be extracted in a scale and rotation invariant manner. Those local features are first computed in both images independently. The two feature sets are then matched in order to establish putative correspondences. Due to the specificity of state-of-the-art feature descriptors such as SIFT [Lowe, 2004] or SURF [Bay et al., 2006], the number of correspondences may already provide a strong indication whether the target object is likely to be contained in the image.

However, there will typically also be a number of mismatches or ambiguous local structures. For this reason, an additional geometric verification stage is applied in order to ensure that the candidate correspondences occur in a consistent geometric configuration.

Thus, the recognition procedure consists of the following basic steps:

---

[1]These are a few global descriptors initially used for specific object recognition. See Section 8.1 of Chapter 8 for an overview of additional window-based representations now typically employed for generic categories.

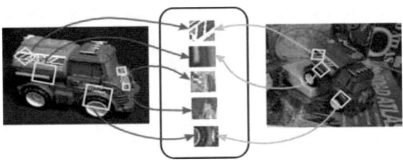

Figure 2.2: Visualization of the local feature-based recognition procedure. Local features (such as SIFT [Lowe, 2004]) are independently extracted from both images, and their descriptors are matched to establish putative correspondences. These candidate matches are then used in order to verify if the features occur in a consistent geometric configuration. Courtesy of David Lowe.

1. *Extract local features* from both the training and test images independently.

2. *Match the feature sets* to find putative correspondences.

3. *Verify* if the matched features occur in a *consistent geometric configuration*.

In the following three chapters (Chapters 3 through 5), we will describe each of these steps in more detail. Then, in Chapter 6, we describe recent systems for specific object recognition that put all of these components together.

CHAPTER 3

Local Features: Detection and Description

Significant progress towards robustly recognizing objects has been made in the past decade through the development of *local invariant features*. These features allow the algorithm to find local image structures in a repeatable fashion and to encode them in a representation that is invariant to a range of image transformations, such as translation, rotation, scaling, and affine deformation. The resulting features then form the basis of approaches for recognizing both specific objects and object categories.

In this chapter, we will explain the basic ideas and implementation steps behind state-of-the-art local feature detectors and descriptors. A more extensive treatment of local features, including detailed comparisons and usage guidelines, can be found in [Tuytelaars and Mikolajczyk, 2007]. Systematic experimental comparisons are reported in [Mikolajczyk and Schmid, 2005, Mikolajczyk et al., 2005].

3.1 INTRODUCTION

The purpose of local invariant features is to provide a representation that allows to efficiently match local structures between images. That is, we want to obtain a sparse set of local measurements that capture the essence of the underlying input images and that encode their *interesting* structure. To meet this goal, the feature extractors must fulfill two important criteria:

- The feature extraction process should be *repeatable and precise*, so that the same features are extracted from two images showing the same object.

- At the same time, the features should be *distinctive*, so that different image structures can be told apart from each other.

In addition, we typically require a sufficient number of feature regions to cover the target object, so that it can still be recognized under partial occlusion. This is achieved by the following feature extraction pipeline, illustrated in Figure 3.1:

1. Find a set of *distinctive keypoints*.

2. Define a region around each keypoint in a *scale- or affine-invariant* manner.

3. Extract and *normalize* the region content.

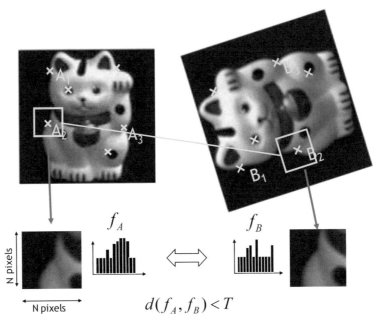

Figure 3.1: An illustration of the recognition procedure with local features. We first find distinctive keypoints in both images. For each such keypoint, we then define a surrounding region in a scale- and rotation-invariant manner. We extract and normalize the region content and compute a local descriptor for each region. Feature matching is then performed by comparing the local descriptors using a suitable similarity measure. Courtesy of Krystian Mikolajczyk.

4. Compute a *descriptor* from the normalized region.

5. Match the local descriptors.

In the remainder of this chapter, we will discuss the keypoint and descriptor steps in detail. Then in the following chapter, we will describe methods for computing candidate matches among the descriptors. Once we have candidate matches, we can then proceed to verify their geometric relationships, as we will describe in Chapter 5.

3.2   DETECTION OF INTEREST POINTS AND REGIONS

3.2.1   KEYPOINT LOCALIZATION

The first step of the local feature extraction pipeline is to find a set of distinctive keypoints that can be reliably localized under varying imaging conditions, viewpoint changes, and in the presence of noise. In particular, the extraction procedure should yield the same feature locations if the input image is translated or rotated. It is obvious that those criteria cannot be met for all image points. For instance, if we consider a point lying in a uniform region, we cannot determine its exact motion, since we cannot distinguish the point from its neighbors. Similarly, if we consider a point on a straight line, we can only measure its motion perpendicular to the line. This motivates us to focus on a particular subset of points, namely those exhibiting signal changes in two directions. In the following, we will present two keypoint detectors that employ different criteria for finding such regions: the *Hessian detector* and the *Harris detector*.

3.2.1.1 The Hessian Detector

The *Hessian detector* [Beaudet, 1978] searches for image locations that exhibit strong derivatives in two orthogonal directions. It is based on the matrix of second derivatives, the so-called Hessian. As derivative operations are sensitive to noise, we always use *Gaussian derivatives* in the following, *i.e.*, we combine the derivative operation with a Gaussian smoothing step with smoothing parameter σ.

$$\mathbf{H}(\mathbf{x}, \sigma) = \left[\begin{array}{cc} I_{xx}(\mathbf{x}, \sigma) & I_{xy}(\mathbf{x}, \sigma) \\ I_{xy}(\mathbf{x}, \sigma) & I_{yy}(\mathbf{x}, \sigma) \end{array} \right]. \tag{3.1}$$

The detector computes the second derivatives I_{xx}, I_{xy}, and I_{yy} for each image point and then searches for points where the determinant of the Hessian becomes maximal:

$$\det(\mathbf{H}) = I_{xx} I_{yy} - I_{xy}^2. \tag{3.2}$$

This search is usually performed by computing a result image containing the Hessian determinant values and then applying *non-maximum suppression* using a 3×3 window. In this procedure, the search window is swept over the entire image, keeping only pixels whose value is larger than the values of all 8 immediate neighbors inside the window. The detector then returns all remaining locations whose value is above a pre-defined threshold θ. As shown in Figure 3.7(top left), the resulting detector responses are mainly located on corners and in strongly textured image areas.

3.2.1.2 The Harris Detector

The popular *Harris/Förstner detector* [Förstner and Gülch, 1987, Harris and Stephens, 1988] was explicitly designed for geometric stability. It defines keypoints to be "points that have locally maximal self-matching precision under translational least-squares template matching" [Triggs, 2004]. In practice, these keypoints often correspond to corner-like structures. The detection procedure is visualized in Figure 3.2.

The Harris detector proceeds by searching for points $\mathbf{x}$ where the second-moment matrix $\mathbf{C}$ around $\mathbf{x}$ has two large eigenvalues. The matrix $\mathbf{C}$ can be computed from the first derivatives in a window around $\mathbf{x}$, weighted by a Gaussian $G(\mathbf{x}, \tilde{\sigma})$:

$$\mathbf{C}(\mathbf{x}, \sigma, \tilde{\sigma}) = G(\mathbf{x}, \tilde{\sigma}) \star \left[\begin{array}{cc} I_x^2(\mathbf{x}, \sigma) & I_x I_y(\mathbf{x}, \sigma) \\ I_x I_y(\mathbf{x}, \sigma) & I_y^2(\mathbf{x}, \sigma) \end{array} \right]. \tag{3.3}$$

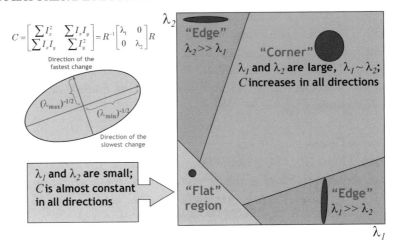

Figure 3.2: The Harris detector searches for image neighborhoods where the second-moment matrix **C** has two large eigenvalues, corresponding to two dominant orientations. The resulting points often correspond to corner-like structures. Courtesy of Dennis Simakov and Darya Frolova.

In this formulation, the convolution with the Gaussian $G(\mathbf{x}, \tilde{\sigma})$ takes the role of summing over all pixels in a circular local neighborhood, where each pixel's contribution is additionally weighted by its proximity to the center point.

Instead of explicitly computing the eigenvalues of **C**, the following equivalences are used

$$\det(\mathbf{C}) = \lambda_1\lambda_2 \tag{3.4}$$
$$\operatorname{trace}(\mathbf{C}) = \lambda_1 + \lambda_2 \tag{3.5}$$

to check if their ratio $r = \frac{\lambda_1}{\lambda_2}$ is below a certain threshold. With

$$\frac{\operatorname{trace}^2(\mathbf{C})}{\det(\mathbf{C})} = \frac{(\lambda_1 + \lambda_2)^2}{\lambda_1\lambda_2} = \frac{(r\lambda_2 + \lambda_2)^2}{r\lambda_2^2} = \frac{(r+1)^2}{r} \tag{3.6}$$

this can be expressed by the following condition

$$\det(\mathbf{C}) - \alpha\operatorname{trace}^2(\mathbf{C}) > t, \tag{3.7}$$

which avoids the need to compute the exact eigenvalues. Typical values for α are in the range of $0.04 - 0.06$. The parameter $\tilde{\sigma}$ is usually set to 2σ, so that the considered image neighborhood is slightly larger than the support of the derivative operator used.

Figure 3.7(top right) shows the results of the Harris detector and compares them to those of the Hessian. As can be seen, the returned locations are slightly different as a result of the changed selection criterion. In general, it can be stated that Harris locations are more specific to corners,

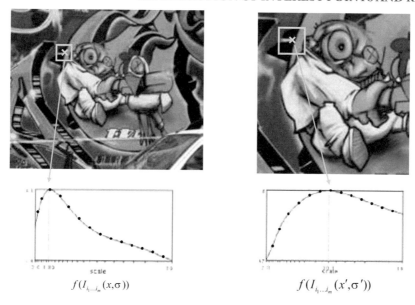

$$f(I_{i_1 \ldots i_m}(x, \sigma))$$ $$f(I_{i_1 \ldots i_m}(x', \sigma'))$$

Figure 3.3: The principle behind automatic scale selection. Given a keypoint location, we evaluate a scale-dependent signature function on the keypoint neighborhood and plot the resulting value as a function of the scale. If the two keypoints correspond to the same structure, then their signature functions will take similar shapes and corresponding neighborhood sizes can be determined by searching for scale-space extrema of the signature function *independently in both images*. Courtesy of Krystian Mikolajczyk.

while the Hessian detector also returns many responses on regions with strong texture variation. In addition, Harris points are typically more precisely located as a result of using first derivatives rather than second derivatives and of taking into account a larger image neighborhood.

Thus, Harris points are preferable when looking for exact corners or when precise localization is required, whereas Hessian points can provide additional locations of interest that result in a denser coverage of the object.

3.2.2   SCALE INVARIANT REGION DETECTION

While shown to be remarkably robust to image plane rotations, illumination changes, and noise [Schmid et al., 2000], the locations returned by the Harris and Hessian detectors are only repeatable up to relatively small scale changes. The reason for this is that both detectors rely on Gaussian derivatives computed at a certain fixed base scale σ. If the image scale differs too much between the test images, then the extracted structures will also be different. For scale invariant feature extraction, it is thus necessary to detect structures that can be reliably extracted under scale changes.

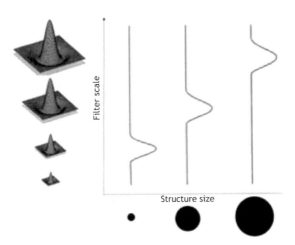

Figure 3.4: The (scale-normalized) Laplacian-of-Gaussian (LoG) is a popular choice for a scale selection filter. Its 2D filter mask takes the shape of a circular center region with positive weights, surrounded by another circular region with negative weights. The filter response is therefore strongest for circular image structures whose radius corresponds to the filter scale.

3.2.2.1 Automatic Scale Selection

The basic idea behind automatic scale selection is visualized in Figure 3.3. Given a keypoint in each image of an image pair, we want to determine whether the surrounding image neighborhoods contain the same structure up to an unknown scale factor. In principle, we could achieve this by sampling each image neighborhood at a range of scales and performing $N \times N$ pairwise comparisons to find the best match. This is, however, too expensive to be of practical use. Instead, we evaluate a *signature function* on each sampled image neighborhood and plot the result value as a function of the neighborhood scale. Since the signature function measures properties of the local image neighborhood at a certain radius, it should take a similar qualitative shape if the two keypoints are centered on corresponding image structures. The only difference will be that one function shape will be squashed or expanded compared to the other as a result of the scaling factor between the two images. Thus, corresponding neighborhood sizes can be detected by searching for extrema of the signature function *independently in both images*. If corresponding extrema σ and σ' are found in both cases, then the scaling factor between the two images can be obtained as $\frac{\sigma'}{\sigma}$.

Effectively, this procedure builds up a *scale space* [Witkin, 1983] of the responses produced by the application of a local kernel with varying scale parameter σ. In order for this idea to work, the signature function or kernel needs to have certain specific properties. It can be shown that the only operator that fulfills all necessary conditions for this purpose is the scale-normalized Gaussian kernel $G(\mathbf{x}, \sigma)$ and its derivatives [Lindeberg, 1994, 1998].

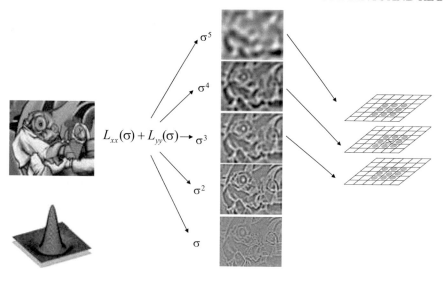

Figure 3.5: The Laplacian-of-Gaussian (LoG) detector searches for 3D scale space extrema of the LoG function. Courtesy of Krystian Mikolajczyk.

3.2.2.2 The Laplacian-of-Gaussian (LoG) Detector

Based on the above idea, Lindeberg proposed a detector for blob-like features that searches for scale space extrema of a scale-normalized *Laplacian-of-Gaussian* (LoG) [Lindeberg, 1998]:

$$L(\mathbf{x}, \sigma) \;=\; \sigma^2 \left(I_{xx}(\mathbf{x}, \sigma) + I_{yy}(\mathbf{x}, \sigma) \right). \tag{3.8}$$

As shown in Figure 3.4, the LoG filter mask corresponds to a circular center-surround structure, with positive weights in the center region and negative weights in the surrounding ring structure. Thus, it will yield maximal responses if applied to an image neighborhood that contains a similar (roughly circular) blob structure at a corresponding scale. By searching for scale-space extrema of the LoG, we can therefore detect circular blob structures.

Note that for such blobs, a repeatable keypoint location can also be defined as the blob center. The LoG can thus both be applied for finding the *characteristic scale* for a given image location and for directly detecting *scale-invariant regions* by searching for 3D (location + scale) extrema of the LoG. This latter procedure is visualized in Figure 3.5 and resulting interest regions are shown in Figure 3.7(bottom left).

3.2.2.3 The Difference-of-Gaussian (DoG) Detector

As shown by Lowe [2004], the scale-space Laplacian can be approximated by a difference-of-Gaussian (DoG) $D(\mathbf{x}, \sigma)$, which can be more efficiently obtained from the difference of two adjacent

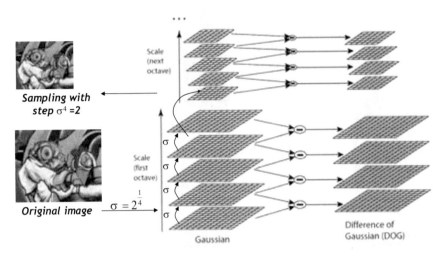

Figure 3.6: The Difference-of-Gaussian (DoG) provides a good approximation for the Laplacian-of-Gaussian. It can be efficiently computed by subtracting adjacent scale levels of a Gaussian pyramid. The DoG region detector then searches for 3D scale space extrema of the DoG function. From Tuytelaars and Mikolajczyk [2007].

scales that are separated by a factor of k:

$$D(\mathbf{x}, \sigma) \quad = \quad (G(\mathbf{x}, k\sigma) - G(\mathbf{x}, \sigma)) \star I(\mathbf{x}) . \qquad (3.9)$$

Lowe [2004] shows that when this factor is constant, the computation already includes the required scale normalization. One can therefore divide each scale octave into an equal number K of intervals, such that $k = 2^{1/K}$ and $\sigma_n = k^n \sigma_0$. For more efficient computation, the resulting scale space can be implemented with a Gaussian pyramid, which resamples the image by a factor of 2 after each scale octave. See Figure 3.6.

As in the case of the LoG detector, DoG interest regions are defined as locations that are simultaneously extrema in the image plane and along the scale coordinate of the $D(\mathbf{x}, \sigma)$ function. Such points are found by comparing the $D(\mathbf{x}, \sigma)$ value of each point with its 8-neighborhood on the same scale level, and with the 9 closest neighbors on each of the two adjacent levels (as depicted in the right side of Figure 3.5).

Since the scale coordinate is only sampled at discrete levels, it is important in both the LoG and the DoG detector to interpolate the responses at neighboring scales in order to increase the accuracy of detected keypoint locations. In the simplest version, this could be done by fitting a second-order polynomial to each candidate point and its two closest neighbors. A more exact approach was introduced by Brown and Lowe [2002]. This approach simultaneously interpolates both the location and scale coordinates of detected peaks by fitting a 3D quadric function.

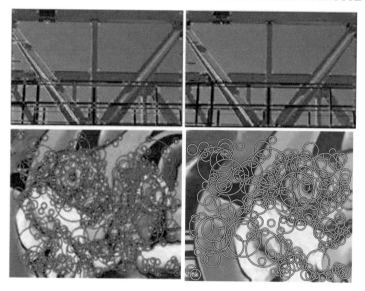

Figure 3.7: Example results of the (top left) Hessian detector; (top right) Harris detector; (bottom left) Laplacian-of-Gaussian detector; (bottom right) Difference-of-Gaussian detector. Courtesy of Krystian Mikolajczyk, from Tuytelaars and Mikolajczyk [2007].

Finally, those regions are kept that pass a threshold t and whose estimated scale falls into a certain scale range $[s_{min}, s_{max}]$. The resulting interest point operator reacts to blob-like structures that have their maximal extent in a radius of approximately 1.6σ of the detected points (as can be derived from the zero crossings of the modeled Laplacian). In order to also capture some of the surrounding structure, the extracted region is typically larger (most current interest region detectors choose a radius of $r = 3\sigma$ around the detected points). Figure 3.7(bottom right) shows the result regions returned by the DoG detector on an example image. It can be seen that the obtained regions are very similar to those of the LoG detector. In practice, the DoG detector is therefore often the preferred choice since it can be computed far more efficiently.

3.2.2.4 The Harris-Laplacian Detector

The Harris-Laplacian operator [Mikolajczyk and Schmid, 2001, 2004] was proposed for increased discriminative power compared to the Laplacian or DoG operators described so far. It combines the Harris operator's specificity for corner-like structures with the scale selection mechanism by Lindeberg [1998]. The method first builds up two separate scale spaces for the Harris function and the Laplacian. It then uses the Harris function to localize candidate points on each scale level and selects those points for which the Laplacian simultaneously attains an extremum over scales.

The resulting points are robust to changes in scale, image rotation, illumination, and camera noise. In addition, they are highly discriminative, as several comparative studies show

[Mikolajczyk and Schmid, 2001, 2003]. As a drawback, however, the original Harris-Laplacian detector typically returns a much smaller number of points than the Laplacian or DoG detectors. This is not a result of changed threshold settings, but of the additional constraint that each point has to fulfill two different maxima conditions simultaneously. For many practical object recognition applications, the lower number of interest regions may be a disadvantage, as it reduces robustness to partial occlusion. This is especially the case for object categorization, where the potential number of corresponding features is further reduced by intra-category variability.

For this reason, an updated version of the Harris-Laplacian detector has been proposed based on a less strict criterion [Mikolajczyk and Schmid, 2004]. Instead of searching for *simultaneous* maxima, it selects scale maxima of the Laplacian at locations for which the Harris function also attains a maximum *at any scale*. As a result, this modified detector yields more interest points at a slightly lower precision, which results in improved performance for applications where a larger absolute number of interest regions is required [Mikolajczyk et al., 2005].

3.2.2.5 The Hessian-Laplace Detector

As in the case of the Harris-Laplace, the same idea can also be applied to the Hessian, leading to the *Hessian-Laplace* detector. As with the single-scale versions, the Hessian-Laplace detector typically returns more interest regions than Harris-Laplace at a slightly lower repeatability [Mikolajczyk et al., 2005].

3.2.3 AFFINE COVARIANT REGION DETECTION

The approaches discussed so far yield local features that can be extracted in a manner that is invariant to translation and scale changes. For many practical problems, however, it also becomes important to find features that can be reliably extracted under large viewpoint changes. If we assume that the scene structure we are interested in is locally planar, then this would boil down to estimating and correcting for the perspective distortion a local image patch undergoes when seen from a different viewpoint. Unfortunately, such a perspective correction is both computationally expensive and error-prone since the local feature patches typically contain only a small number of pixels. It has, however, been shown by a number of researchers [Matas et al., 2002, Mikolajczyk and Schmid, 2004, Schaffalitzky and Zisserman, 2002, Tuytelaars and Van Gool, 2000, 2004] that a local affine approximation is sufficient in such cases.

We therefore aim to extend the region extraction procedure to *affine covariant* regions[1]. While a scale- and rotation-invariant region can be described by a circle, an affine deformation transforms this circle to an ellipse. We thus aim to find local regions for which such an ellipse can be reliably and repeatedly extracted purely from local image properties.

---

[1]The literature speaks of affine *covariant* extraction here in order to emphasize the property that extracted region *shapes* vary according to the underlying affine deformation. This is required so that the region *content* will be *invariant*.

3.2.3.1 Harris and Hessian Affine Detectors

Both the Harris-Laplace and Hessian-Laplace detectors can be extended to yield affine covariant regions. This is done by the following iterative estimation scheme. The procedure is initialized with a circular region returned by the original scale-invariant detector. In each iteration, we build up the region's second-moment matrix and compute the eigenvalues of this matrix. This yields an elliptical shape (as shown in Figure 3.2), corresponding to a local affine deformation. We then transform the image neighborhood such that this ellipse is transformed to a circle and update the location and scale estimate in the transformed image. This procedure is repeated until the eigenvalues of the second-moment matrix are approximately equal.

As a result of this iterative estimation scheme, we obtain a set of elliptical regions which are adapted to the local intensity patterns, so that the same object structures are covered despite the deformations caused by viewpoint changes.

3.2.3.2 Maximally Stable Extremal Regions (MSER)

A different approach for finding affine covariant regions has been proposed by Matas et al. [2002]. In contrast to the above methods, which start from keypoints and progressively add invariance levels, this approach starts from a segmentation perspective. It applies a watershed segmentation algorithm to the image and extracts homogeneous intensity regions which are stable over a large range of thresholds, thus ending up with *Maximally Stable Extremal Regions* (MSER). By construction, those regions are stable over a range of imaging conditions and can still be reliably extracted under viewpoint changes. Since they are generated by a segmentation process, they are not restricted to elliptical shapes, but can have complicated contours. In fact, the contour shape itself is often a good feature, which has led to the construction of specialized contour feature descriptors [Matas et al., 2002]. For consistency with the other feature extraction steps discussed here, an elliptical region can however also easily be fitted to the Maximally Stable regions by computing the eigenvectors of their second-moment matrices.

3.2.3.3 Other Interest Region Detectors

Several other interest region detectors have been proposed that are not discussed here. Tuytelaars & Van Gool introduced detectors for affine covariant *Intensity Based Regions* (IBR) and *Edge Based Regions* (EBR) [Tuytelaars and Van Gool, 2004]. Kadir & Brady proposed a *Salient Regions* detector that was later on also extended to affine covariant extraction [Kadir and Brady, 2001, Kadir et al., 2004]. An overview over those detectors and a discussion of their merits can be found in [Tuytelaars and Mikolajczyk, 2007].

3.2.4 ORIENTATION NORMALIZATION

After a scale-invariant region has been detected, its content needs to be normalized for rotation invariance. This is typically done by finding the region's *dominant orientation* and then rotating the region content according to this angle in order to bring the region into a canonical orientation.

Lowe [2004] suggests the following procedure for the orientation normalization step. For each detected interest region, the region's scale is used to select the closest level of the Gaussian pyramid, so that all following computations are performed in a scale invariant manner. We then build up a gradient orientation histogram with 36 bins covering the 360° range of orientations. For each pixel in the region, the corresponding gradient orientation is entered into the histogram, weighted by the pixel's gradient magnitude and by a Gaussian window centered on the keypoint with a scale of 1.5σ. The highest peak in the orientation histogram is taken as the dominant orientation, and a parabola is fitted to the 3 adjacent histogram values to interpolate the peak position for better accuracy.

In practice, it may happen that multiple equally strong orientations are found for a single interest region. In such cases, selecting only one of them would endanger the recognition procedure since small changes in the image signal could cause one of the other orientations to be chosen instead, which could lead to failed matches. For this reason, Lowe suggests to create a separate interest region for each orientation peak that reaches at least 80% of the dominant peak's value [Lowe, 2004]. This strategy significantly improves the region detector's repeatability at a relatively small additional cost (according to Lowe [2004], only about 15% of the points are assigned multiple orientations).

3.2.5 SUMMARY OF LOCAL DETECTORS

Summarizing the above, we have seen the following local feature detectors so far. If precisely localized points are of interest, we can use the *Harris* and *Hessian* detectors. When looking for scale-invariant regions, we can choose between the *LoG* or *DoG* detectors, both of which react to blob-shaped structures. In addition, we can combine the Harris and Hessian point detectors with the Laplacian scale selection idea to obtain the *Harris-Laplacian* and *Hessian-Laplacian* detectors. Finally, we can further generalize those detectors to affine covariant region extraction, resulting in the *Harris-Affine* and *Hessian-Affine* detectors. The affine covariant region detectors are complemented by the *MSER* detector, which is based on maximally stable segmentation regions. All of those detectors have been used in practical applications. Detailed experimental comparisons can be found in [Mikolajczyk and Schmid, 2004, Tuytelaars and Mikolajczyk, 2007].

3.3 LOCAL DESCRIPTORS

Once a set of interest regions has been extracted from an image, their content needs to be encoded in a descriptor that is suitable for discriminative matching. The most popular choice for this step is the SIFT descriptor [Lowe, 2004], which we present in detail in the following.

3.3.1 THE SIFT DESCRIPTOR

The *Scale Invariant Feature Transform* (SIFT) was originally introduced by Lowe as combination of a DoG interest region detector and a corresponding feature descriptor [Lowe, 1999, 2004]. However, both components have since then also been used in isolation. In particular, a series of studies has

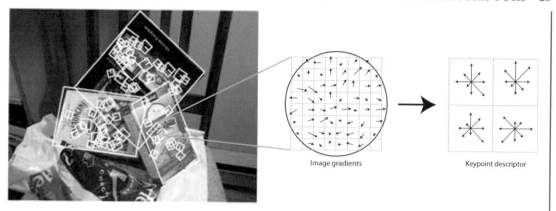

Image gradients   Keypoint descriptor

Figure 3.8: Visualization of the SIFT descriptor computation. For each (orientation-normalized) scale invariant region, image gradients are sampled in a regular grid and are then entered into a larger 4 × 4 grid of local gradient orientation histograms (for visibility reasons, only a 2 × 2 grid is shown here). Based on Lowe [1999] and Lowe [2004].

confirmed that the SIFT descriptor is suitable for combination with all of the above-mentioned region detectors and that it achieves generally good performance [Mikolajczyk and Schmid, 2005].

The SIFT descriptor aims to achieve robustness to lighting variations and small positional shifts by encoding the image information in a *localized set of gradient orientation histograms*. The descriptor computation starts from a scale and rotation normalized region extracted with one of the above-mentioned detectors. As a first step, the image gradient magnitude and orientation is sampled around the keypoint location using the region scale to select the level of Gaussian blur (*i.e.*, the level of the Gaussian pyramid at which this computation is performed). Sampling is performed in a regular grid of 16 × 16 locations covering the interest region. For each sampled location, the gradient orientation is entered into a coarser 4 × 4 grid of gradient orientation histograms with 8 orientation bins each, weighted by the corresponding pixel's gradient magnitude and by a circular Gaussian weighting function with a σ of half the region size. The purpose of this Gaussian window is to give higher weights to pixels closer to the middle of the region, which are less affected by small localization inaccuracies of the interest region detector.

This procedure is visualized for a smaller 2 × 2 grid in Figure 3.8. The motivation for this choice of representation is that the coarse spatial binning allows for small shifts due to registration errors without overly affecting the descriptor. At the same time, the high-dimensional representation provides enough discriminative power to reliably distinguish a large number of keypoints.

When computing the descriptor, it is important to avoid all boundary effects, both with respect to spatial shifts and to small orientation changes. Thus, when entering a sampled pixel's gradient information into the 3-dimensional spatial/orientation histogram, its contribution should be smoothly distributed among the adjoining histogram bins using trilinear interpolation.

Once all orientation histogram entries have been completed, those entries are concatenated to form a single $4 \times 4 \times 8 = 128$ dimensional feature vector. A final illumination normalization completes the extraction procedure. For this, the vector is first normalized to unit length, thus adjusting for changing image contrast. Then all feature dimensions are thresholded to a maximum value of 0.2 and the vector is again normalized to unit length. This last step compensates for non-linear illumination changes due to camera saturation or similar effects.

3.3.2   THE SURF DETECTOR/DESCRIPTOR

As local feature detectors and descriptors have become more widespread, efficient implementations are getting more and more important. Consequently, several approaches have been proposed in order to speed up the interest region extraction and/or description stages [Bay et al., 2008, 2006, Cornelis and Van Gool, 2008, Rosten and Drummond, 2008]. Among those, we want to pick out the SURF ("Speeded-Up Robust Features") approach, which has been designed as an efficient alternative to SIFT [Bay et al., 2008, 2006].

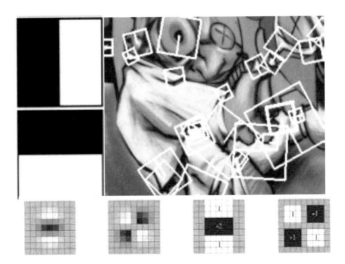

Figure 3.9: The SURF detector and descriptor were designed as an efficient alternative to SIFT. Instead of relying on ideal Gaussian derivatives, their computation is based on simple 2D box filters, which can be efficiently evaluated using integral images. Based on Bay et al. [2008].

SURF combines a Hessian-Laplace region detector with its own gradient orientation-based feature descriptor. Instead of relying on Gaussian derivatives for its internal computations, it is however based on simple 2D box filters ("Haar wavelets"), as shown in Figure 3.9. Those box filters approximate the effects of the derivative filter kernels, but they can be efficiently evaluated using integral images [Viola and Jones, 2004]. In particular, this evaluation requires the same constant number of lookups regardless of the image scale, thus removing the need for a Gaussian pyramid.

Despite this simplification, SURF has been shown to achieve comparable repeatability as detectors based on standard Gaussian derivatives, while yielding speedups of more than a factor of five compared to standard DoG.

The SURF descriptor is also motivated by SIFT and pursues a similar spatial binning strategy, dividing the feature region into a 4×4 grid. However, instead of building up a gradient orientation histogram for each bin, SURF only computes a set of summary statistics $\sum dx$, $\sum |dx|$, $\sum dy$, and $\sum |dy|$, resulting in a 64-dimensional descriptor, or a slightly extended set resulting in a 128-dimensional descriptor version.

Motivated by the success of SURF, a further optimized version has been proposed by Cornelis and Van Gool [2008] that takes advantage of the computational power available in current CUDA enabled graphics cards. This GPUSURF implementation has been reported to perform feature extraction for a 640×480 image at frame rates up to 200 Hz (*i.e.*, taking only 5ms per frame), thus making feature extraction a truly affordable processing step.

3.4   CONCLUDING REMARKS

The first step in the specific-object recognition pipeline is to extract local features from both the training and test images. For this, we can use any of the feature detectors and descriptors described in this chapter. The appropriate choice of detector class (single-scale, scale-invariant, or affine-invariant) mainly depends on the type of viewpoint variations foreseen for the target application.

For many practical recognition applications, scale invariant features (in particular SIFT [Lowe, 2004]) have proven a good compromise, since they are fast to extract, are robust to moderate viewpoint variations, and retain enough discriminative power to allow for reliable matching. When dealing with larger viewpoint changes, as in wide-baseline stereo matching applications, affine invariance becomes important in order to still establish correspondences. However, the added invariance comes at the price of reduced discriminative power (since several different elliptical regions can be mapped onto the same circular descriptor neighborhood) and generally also a smaller number of features (since not all regions have a characteristic affine-covariant neighborhood). Whenever possible, it is therefore advisable to use the simpler feature representation.

The development of local invariant features has had an enormous impact in many areas of computer vision, including wide-baseline stereo matching, image retrieval, object recognition, and categorization. They have provided the basis for many state-of-the-art algorithms and have led to a number of new developments. Moreover, efficient implementations for all detectors discussed in this chapter are freely available [GPU, 2008, Oxf, 2004, SUR, 2006], making them truly building blocks that other researchers can build on.

In the next two chapters, we discuss how to compute candidate matching descriptors, and then describe how the geometric consistency of those candidate matches are verified to perform specific object recognition. In later chapters we will again draw on local descriptors to build models for generic object categorization.

CHAPTER 4

Matching Local Features

In the previous chapter, we saw several specific techniques to detect repeatable interest points in an image (Section 3.2) and then robustly describe the local appearance at each such point (Section 3.3). Now, given an image and its local features, we need to be able to *match* them to similar-looking local features in other images (e.g., to model images of the specific objects we are trying to recognize). See Figure 4.1. To identify candidate matches, we essentially want to search among all previously seen local descriptors, and retrieve those that are nearest according to Euclidean distance in the feature space (such as the 128-dimensional "SIFT space").

Because the local descriptions are by design invariant to rotations, translations, scalings, and some photometric effects, this matching stage will be able to tolerate reasonable variations in viewpoint, pose, and illumination across the views of the object. Further, due to the features' distinctiveness, if we detect a good correspondence based on the local feature matches alone, we will already have a reasonable measure of how likely it is that two images share the same object. However, to strengthen confidence and eliminate ambiguous matches, it is common to follow the matching process discussed in this chapter with a check for geometric consistency, as we will discuss in Chapter 5.

The naive solution to identifying local feature matches is straightforward: simply scan through all previously seen descriptors, compare them to the current input descriptor, and take those within some threshold as candidates. Unfortunately, however, such a linear-time scan is usually unrealistic in terms of computational complexity. In many practical applications, one has to search for matches in a database of millions of features. Thus, efficient algorithms for *nearest neighbor* or *similarity search* are crucial. The focus of this chapter is to describe the algorithms frequently used in the recognition pipeline to rapidly match local descriptors.[1]

Specifically, in the first section of this chapter we overview both tree-based algorithms for exact near neighbor search, as well as approximate nearest neighbor algorithms (largely hashing-based) that are more amenable for high-dimensional descriptors. Then, in Section 4.2, we describe a frequently used alternative based on *visual vocabularies*. Instead of performing similarity search in the "raw" vector space of the descriptors, the vocabulary-based method first quantizes the feature space into discrete "visual words", making it possible to index feature matches easily with an inverted file.

---

[1]In fact, while our focus at this stage is on local feature matching for specific object recognition, most of the algorithms discussed are quite general and also come into play for other recognition-related search tasks, such as near-neighbor image retrieval.

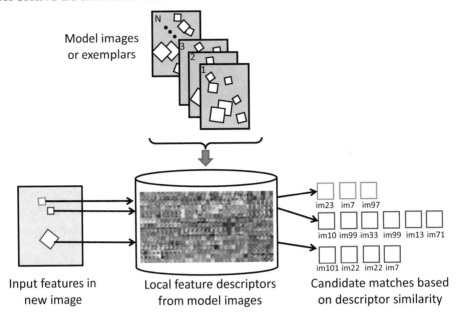

Model images
or exemplars

Input features in
new image

Local feature descriptors
from model images

Candidate matches based
on descriptor similarity

im23 im7 im97

im10 im99 im33 im9 im13 im71

im101 im22 im22 im7

Figure 4.1: The goal when matching local features is to find those descriptors from any previously seen model (exemplar) that are near in the feature space to those local features in a novel image (depicted on the left). Since each exemplar image may easily contain on the order of hundreds to thousands of interest points, the database of descriptors quickly becomes very large; to make searching for matches practical, the database must be mapped to data structures for efficient similarity search.

4.1 EFFICIENT SIMILARITY SEARCH

What methods are effective for retrieving descriptors relevant to a new image? The choice first depends on the dimensionality of the descriptors. For low-dimensional points, effective data structures for exact nearest neighbor search are known, *e.g.*, *kd*-trees [Friedman et al., 1977]. For high-dimensional points, these methods become inefficient, and so researchers often employ *approximate* similarity search methods. This section overviews examples of both such techniques that are widely used in specific-object matching.

4.1.1 TREE-BASED ALGORITHMS

Data structures using spatial partitions and recursive hyperplane decomposition provide an efficient means to search low-dimensional vector data exactly. The *kd*-tree [Friedman et al., 1977] is one such approach that has often been employed to match local descriptors, in several variants (e.g., [Beis and Lowe, 1997, Lowe, 2004, Muja and Lowe, 2009, Silpa-Anan and Hartley, 2008]). The *kd*-tree is a binary tree storing a database of *k*-dimensional points in its leaf nodes. It recursively partitions the points into axis-aligned cells, dividing the points approximately in half by a line perpendicular to one of the k coordinate axes. The division strategies aim to maintain balanced trees and/or uniformly shaped cells—for example, by choosing the next axis to split according to that which has the largest variance among the database points, or by cycling through the axes in order.

To find the point nearest to some query, one traverses the tree following the same divisions that were used to enter the database points; upon reaching a leaf node, the points found there are compared to the query. The nearest one becomes the "current best". While the point is nearer than others to the query, it need not be the absolute nearest (for example, consider a query occurring near the initial dividing split at the top of the tree, which can easily be nearer to some points on the other side of the dividing hyperplane). Thus, the search continues by backtracking along the unexplored branches, checking whether the circle formed about the query by the radius given by the current best match intersects with a subtree's cell area. If it does, that subtree is considered further, and any nearer points found as the search recurses are used to update the current best. If not, the subtree can be pruned. See Figure 4.2 for a sketch of an example tree and query.

The procedure guarantees that the nearest point will be found.[2] Constructing the tree for N database points (an offline cost for a single database) requires $O(N \log N)$ time. Inserting points requires $O(\log N)$ time. Processing a query requires $O(N^{1-\frac{1}{k}})$ time, and the algorithm is known to be quite effective for low-dimensional data (i.e., fewer than 10 dimensions).

In high-dimensional spaces, however, the algorithm ends up needing to visit many more branches during the backtracking stage, and in general degrades to worst case linear scan performance in practice. The particular behavior depends not only on the dimension of the points, but also the distribution of the database examples that have been indexed, combined with the choices in how divisions are computed.

Other types of tree data structures can operate with arbitrary *metrics* [Ciaccia et al., 1997, Uhlmann, 1991], removing the requirement of having data in a vector space by exploiting the triangle inequality. However, similar to kd-trees, the metric trees in practice rely on good heuristics for selecting useful partitioning strategies, and, in spite of logarithmic query times in the expectation, also degenerate to a linear time scan of all items depending on the distribution of distances for the data set.

Since high-dimensional image descriptors are commonly used in object recognition, several strategies to mitigate these factors have been explored. One idea is to relax the search requirement to allow the return of *approximate* nearest neighbors, using a variant of kd-trees together with a priority queue [Arya et al., 1998, Beis and Lowe, 1997]. Another idea is to generate *multiple* randomized kd-trees (e.g., by sampling splits according to the coordinates' variance), and then process the query in all trees using a single priority queue across them [Silpa-Anan and Hartley, 2008]. Given the sensitivity of the algorithms to the data distribution, some recent work also attempts to automatically select algorithm configurations and parameters for satisfactory performance by using a cross-validation approach [Muja and Lowe, 2009]. Another interesting direction pursued for improving the efficiency and effectiveness of tree-based search involves integrating learning or the matching task into the

---

[2]Range queries and k-nn queries are also supported.

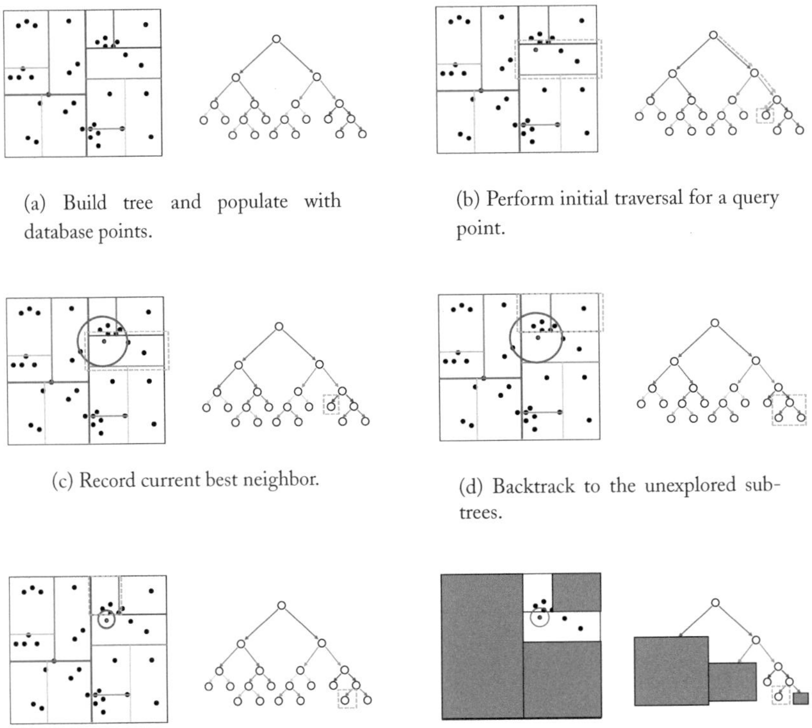

(a) Build tree and populate with database points.

(b) Perform initial traversal for a query point.

(c) Record current best neighbor.

(d) Backtrack to the unexplored subtrees.

(e) Update nearest point and distance if a closer point is found.

(f) Disregard subtrees that cannot be any closer.

Figure 4.2: Sketch of kd-tree processing. (a) The database points are entered into a binary tree, where each division is an axis-aligned hyperplane. (b) Given a new query (red point) for which we wish to retrieve the nearest neighbor, first the tree is traversed, choosing the left or right subtree at each node according to the query's value in the coordinate along which this division was keyed. The green dotted box denotes the cell containing the points in the leaf node reached by this query. (c) At this point, we know the current best point is that in the leaf node that is closest to the query, denoted with the outer red circle. (d) Then we backtrack and consider the other branch at each node that was visited, checking if its cell intersects the "current best" circle around the query. (e) If so, its subtree is explored further, and the current best radius is updated if a nearer point is found. (f) Continue, and prune subtrees once a comparison at its root shows that it cannot improve on the current nearest point. Courtesy of The Auton Lab, Carnegie Mellon University.

tree construction, for example by using decision trees in which each internal node is associated with a weak classifier built with simple measurements from the feature patches [Lepetit et al., 2005, Obdrzalek and Matas, 2005].

4.1.2   HASHING-BASED ALGORITHMS AND BINARY CODES

Hashing algorithms are an effective alternative to tree-based data structures. Motivated by the inadequacy of existing *exact* nearest-neighbor techniques to provide sub-linear time search for high-dimensional data (including the kd-tree and metric tree approaches discussed above), randomized *approximate* hashing-based similarity search algorithms have been explored. The idea in approximate similarity search is to trade off some precision in the search for the sake of substantial query time reductions. More specifically, guarantees are of the general form: if for a query point q there exists a database point x such that $d(q, x) \leq r$ for some search radius r, then, with high probability a point x' is returned such that $d(q, x') \leq (1 + \epsilon)r$. Otherwise, the absence of such a point is reported.[3]

4.1.2.1   Locality Sensitive Hashing

Locality-sensitive hashing (LSH) [Charikar, 2002, Datar et al., 2004, Gionis et al., 1999, Indyk and Motwani, 1998] is one such algorithm that offers sub-linear time search by hashing highly similar examples together in a hash table. The idea is that if one can guarantee that a randomized hash function will map two inputs to the same bucket with high probability only if they are similar, then, given a new query, one needs only to search the colliding database examples to find those that are most probable to lie in the input's near neighborhood.[4] The search is approximate, however, and one sacrifices a predictable degree of error in the search in exchange for a significant improvement in query time.

More formally, a family of LSH functions $\mathcal{F}$ is a distribution of functions where for any two objects x_i and x_j,

$$\Pr_{h \in \mathcal{F}} \left[h(x_i) = h(x_j) \right] = sim(x_i, x_j), \tag{4.1}$$

where $sim(x_i, x_j) \in [0, 1]$ is some similarity function, and $h(x)$ is a hash function drawn from $\mathcal{F}$ that returns a single bit [Charikar, 2002]. Concatenating a series of b hash functions drawn from $\mathcal{F}$ yields b-dimensional hash keys. When $h(x_i) = h(x_j)$, x_i and x_j collide in the hash table. Because the probability that two inputs collide is equal to the similarity between them, highly similar objects are indexed together in the hash table with high probability. On the other hand, if two objects are very dissimilar, they are unlikely to share a hash key (see Figure 4.3). At query time, one maps the query to its hash bucket, pulls up any database instances also in that bucket, and then exhaustively searches only those (few) examples. In practice, multiple hash tables are often used, each with independently

---

[3]Variants of this guarantee for nearest neighbors (rather than r-radius neighbors) also exist.

[4]Locality sensitive hashing has been formulated in two related contexts—one in which the likelihood of collision is guaranteed relative to a threshold on the radius surrounding a query point [Indyk and Motwani, 1998], and another where collision probabilities are equated with a similarity function score [Charikar, 2002]. We use the latter definition here.

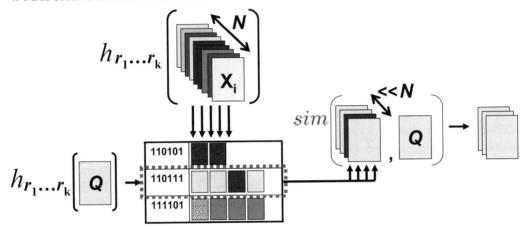

Figure 4.3: Overview of locality-sensitive hashing. If hash functions guarantee a high probability of collision for features that are similar under a metric of interest, one can search a large database in sub-linear time via locality sensitive hashing techniques [Charikar, 2002, Indyk and Motwani, 1998]. A list of k hash functions $h_{r_1}, \ldots, h_{r_k}$ are applied to map N database images to a hash table where similar items are likely to share a bucket. After hashing a query $\mathbf{Q}$, one must only evaluate the similarity between $\mathbf{Q}$ and the database examples with which it collides to obtain the approximate near-neighbors. From Kulis et al. [2009]. Copyright © 2009 IEEE.

drawn hash functions, and the query is compared against the union of the database points to which it hashes in all tables.

Given valid LSH functions, the query time for retrieving $(1 + \epsilon)$-near neighbors is bounded by $O(N^{1/(1+\epsilon)})$ for the Hamming distance and a database of size N [Gionis et al., 1999]. One can therefore trade off the accuracy of the search with the query time required. Early LSH functions were developed to accommodate the Hamming distance [Indyk and Motwani, 1998], inner products [Charikar, 2002], and ℓ_p norms [Datar et al., 2004]. These methods were quickly adopted by vision researchers for a variety of image search applications [Shakhnarovich et al., 2006].

Since meaningful image comparisons for recognition often demand richer comparison measures, work in the vision community has developed novel locality-sensitive hash functions for additional classes of metrics. For example, an embedding of the normalized partial matching between two sets of local features is given in [Grauman and Darrell, 2007a] that allows sub-linear time hashing for the pyramid match kernel (see Section 10.2.1.1 below). A related form of hashing computes *sketches* of feature sets and allows search according to the sets' overlap [Broder, 1998]; this "Min-Hash" framework has been demonstrated and extended for near-duplicate detection and image search in [Chum et al., 2008]. Most recently, a kernelized form of LSH (KLSH) is proposed in [Kulis and Grauman, 2009], which makes it possible to perform locality-sensitive hashing for arbitrary kernel functions. Results are shown for various kernels relevant to object recognition, including the χ^2 kernel that is often employed for comparing bag-of-words descriptors (to be defined below).

Aside from widening the class of metrics and kernels supportable with LSH, researchers have also considered how to integrate machine learning elements so that the hash functions are better suited for a particular task. For object recognition, this means that one wants hash functions that are more likely to map instances of the same object to the same hash buckets, or, similarly, patch descriptors from the same real-world object point to the same bucket. For example, Parameter Sensitive Hashing (PSH) [Shakhnarovich et al., 2003] is an LSH-based algorithm that uses boosting to select hash functions that best reflect similarity in a parameter space of interest. *Semi-supervised hash functions* make it possible to efficiently index data according to learned distances [Jain et al., 2008a, Kulis et al., 2009, Strecha et. al., 2010, Wang, Kumar and Chang, 2010]. Typically, supervision is given in the form of similar and dissimilar pairs of instances, and then while the metric learning algorithm updates its parameters to best capture those constraints, the hash functions' parameters are simultaneously adjusted. While most methods assume that all supervision is available in "batch" at the onset, online metric learners that accumulate constraints over time together with hash tables that can be updated incrementally have also been developed [Jain et al., 2008b].

4.1.2.2 Binary Embedding Functions

Embedding functions are a related mechanism that are used to map expensive distance functions into something more manageable computationally. Either constructed or learned, these embeddings aim to approximately preserve the desired distance function when mapping to a low-dimensional space that is more easily searchable with known techniques. Informally, given an original feature space $\mathcal{X}$ and associated distance function $d_{\mathcal{X}}$, the basic idea is to designate a function $f : \mathcal{X} \to \mathcal{E}$ that maps the inputs into a new space $\mathcal{E}$ with associated distance $d_{\mathcal{E}}$ in such a way that $d_{\mathcal{E}}(f(\boldsymbol{x}), f(\boldsymbol{y})) \approx d_{\mathcal{X}}(\boldsymbol{x}, \boldsymbol{y})$, for any $\boldsymbol{x}, \boldsymbol{y} \in \mathcal{X}$. Often the target space for the embedding is the Hamming space. Such binary codes have the advantage of requiring minimal memory; they also permit fast bit counting routines for the Hamming distance, and can be indexed directly using the computer's memory addresses.

Work in the vision and learning community has developed useful embedding functions that aim to preserve a variety of similarity metrics with simple low-dimensional binary codes. For example, the BoostMap [Athitsos et al., 2004] and Boosted Similarity Sensitive Coding (Boost-SSC) [Shakhnarovich, 2005] algorithms learn an embedding using different forms of boosting, combining multiple weighted 1D embeddings so as to preserve the proximity structure given by the original distance function. Building on this notion, more recent work develops Semantic Hashing algorithms that train embedding functions using boosting or multiple layers of restricted Boltzmann machines [Salakhutdinov and Hinton, 2007, Torralba et al., 2008]; results show the impact for searching Gist image descriptors [Torralba et al., 2008]. Embeddings based on random projections have also been explored for shift-invariant kernels, which includes a Gaussian kernel [Raginsky and Lazebnik, 2009].

Such methods are related to LSH in the sense that both seek small "keys" that can be used to encode similar inputs, and often these keys exist in Hamming space. However, note that while hash functions also typically map the data to binary strings (the "hash keys"), in that case the codes serve to insert instances into buckets, whereas technically the embedding function outputs are treated as a new feature space in which to perform the similarity search.

4.1.3 A RULE OF THUMB FOR REDUCING AMBIGUOUS MATCHES

When matching local feature sets extracted from real-world images, many features will stem from background clutter and will therefore have no meaningful neighbor in the other set. Other features lie on repetitive structures and may therefore have ambiguous matches (for example, imagine an image containing a building with many identical windows). Hence, one needs to find a way to distinguish reliable matches from unreliable ones. This cannot be done based on the descriptor distance alone, since some descriptors are more discriminative than others.

An often-used strategy (initially proposed by Lowe [2004]) is to consider the ratio of the distance to the closest neighbor to that of the second-closest one as a decision criterion. Specifically, we identify the nearest neighbor local feature originating from an exemplar in the database of training images, and then consider the second nearest neighbor that originates from a different object than the nearest neighbor feature. If the ratio of the distance to the first neighbor over the distance to the second neighbor is relatively large, this is a sign that the match may be ambiguous. Similarly, if the ratio is low, it suggests that it is a reliable match. This strategy effectively penalizes features that come from a densely populated region of feature space and that are therefore more ambiguous. By comparing the probability density functions of correct and incorrect matches in quantitative experiments, Lowe arrives at the recommendation to reject all matches in which the distance ratio is greater than 0.8, which in his experiments eliminated 90% of the false matches while discarding less than 5% correct matches [Lowe, 2004].

4.2 INDEXING FEATURES WITH VISUAL VOCABULARIES

In this section, we overview the concept of a *visual vocabulary*—a strategy that draws inspiration from the text retrieval community and enables efficient indexing for local image features. Rather than preparing a tree or hashing data structure to aid in direct similarity search, the idea is to *quantize* the local feature space. By mapping the local descriptors to discrete tokens, we can then "match" them by simply looking up features assigned to the identical token.

In the following, we first describe the formation of visual words (Sections 4.2.1 through 4.2.3), and then describe their utility for indexing (Section 4.2.4). Note that we will return to this representation later in Section 8.1 in the context of object categorization, as it is the basis for the simple but effective "bag-of-words" image descriptor.

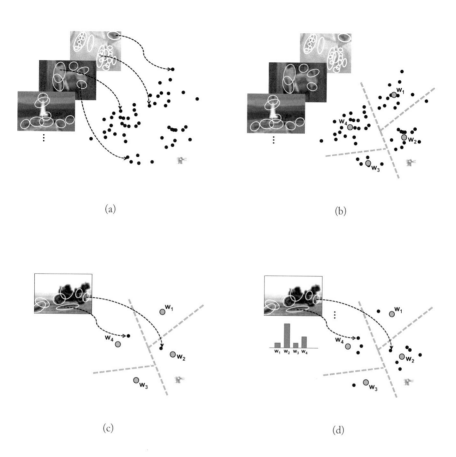

(a)   (b)

(c)   (d)

Figure 4.4: A schematic to illustrate visual vocabulary construction and word assignment. (a) A large corpus of representative images are used to populate the feature space with descriptor instances. The white ellipses denote local feature regions, and the black dots denote points in some feature space, *e.g.*, SIFT. (b) Next, the sampled features are clustered in order to quantize the space into a discrete number of visual words. The visual words are the cluster centers, denoted with the large green circles. The dotted green lines signify the implied Voronoi cells based on the selected word centers. (c) Now, given a new image, the nearest visual word is identified for each of its features. This maps the image from a set of high-dimensional descriptors to a list of word numbers. (d) A bag-of-visual-words histogram can be used to summarize the entire image (see Section 8.1). It counts how many times each of the visual words occurs in the image.

4.2.1 CREATING A VISUAL VOCABULARY

Methods for indexing and efficient retrieval with text documents are mature, and effective enough to operate with millions or billions of documents at once [Baeza-Yates and Ribeiro-Neto, 1999]. Documents of text contain some distribution of words, and thus can be compactly summarized by their word counts (known as a bag-of-words). Since the occurrence of a given word tends to be sparse across different documents, an index that maps words to the files in which they occur can map a keyword query directly to potentially relevant content. For example, if we query a document database with the word "car", we should immediately eliminate the many documents that never mention the word "car".

What cues, then, can one take from text processing to aid visual search? An image is a sort of document, and (using the representations introduced in Chapter 3) it contains a set of local feature descriptors. However, at first glance, the analogy would stop there: text words are discrete "tokens", whereas local image descriptors are high-dimensional, real-valued feature points. How could one obtain discrete "visual words"?

To do so, we must impose a quantization on the feature space of local image descriptors. That way, any novel descriptor vector can be coded in terms of the (discretized) region of feature space to which it belongs. The standard pipeline to form a so-called "visual vocabulary" consists of (1) collecting a large sample of features from a representative corpus of images and (2) quantizing the feature space according to their statistics. Often simple *k-means clustering* is used to perform the quantization; one initializes the k cluster centers with randomly selected features in the corpus, and then iterates between updating each point's cluster membership (based on which cluster center it is nearest to) and updating the k means (based on the mean of the points previously assigned to each cluster). In that case, the visual "words" are the k cluster centers, and the size of the vocabulary k is a user-supplied parameter. Once the vocabulary is established, the corpus of sampled features can be discarded. Then a novel image's features can be translated into words by determining which visual word they are nearest to in the feature space (*i.e.*, based on the Euclidean distance between the cluster centers and the input descriptor). See Figure 4.4 for a diagram of the procedure.

Drawing inspiration from text retrieval methods, Sivic and Zisserman proposed quantizing local image descriptors for the sake of rapidly indexing video frames with an inverted file [Sivic and Zisserman, 2003]. They showed that local descriptors extracted at interest points could be mapped to visual words by computing prototypical descriptors with k-means clustering, and that having these tokens enabled faster retrieval of frames containing the same words. Furthermore, they showed the potential of exploiting a *term frequency-inverse document frequency* weighting on the words, which de-emphasizes those words that are common to many images and thus possibly less informative, and a *stop-list*, which ignores extremely frequent words that appear in nearly every image (analogous to "a" or "the" in text).

What will a visual word capture? The answer depends on several factors, including what corpus of features is used to build the vocabulary, the number of words selected, the quantization algorithm used, and the interest point or sampling mechanism chosen for feature extraction. Intuitively, the

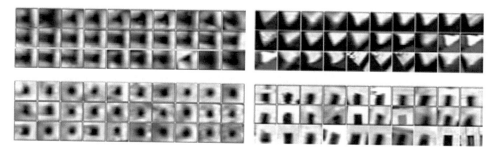

Figure 4.5: Four examples of visual words. Each group shows instances of patches that are assigned to the same visual word. From Sivic and Zisserman [2003]. Copyright © 2003 IEEE.

larger the vocabulary, the more fine-grained the visual words. In general, patches assigned to the same visual word should have similar low-level appearance (see Figure 4.5). Particularly when the vocabulary is formed in an unsupervised manner, there are no constraints that the common types of local patterns be correlated with object-level parts. However, in later chapters we will see some methods that use visual vocabularies or codebooks to provide candidate parts to a part-based category model.

4.2.2   VOCABULARY TREES

The discussion above assumes a flat quantization of the feature space, but many current techniques exploit hierarchical partitions [Bosch et al., 2007a, Grauman and Darrell, 2005, 2006a, Moosmann et al., 2006, Nister and Stewenius, 2006, Yeh et al., 2007]. In particular, the *vocabulary tree* approach [Nister and Stewenius, 2006] uses hierarchical k-means to recursively subdivide the feature space, given a choice of the branching factor and number of levels. Vocabulary trees offer a significant advantage in terms of the computational cost of assigning novel image features to words—from linear to logarithmic in the size of the vocabulary. This in turn makes it practical to use much larger vocabularies (*e.g.*, on the order of one million words).

   Experimental results suggest that these more specific words (smaller quantized bins) are particularly useful for matching features for specific instances of objects [Nister and Stewenius, 2006, Philbin et al., 2007, 2008]. Since quantization entails a hard-partitioning of the feature space, it can also be useful in practice to use multiple randomized hierarchical partitions, and/or to perform a soft assignment in which a feature results in multiple weighted entries in nearby bins.

4.2.3   CHOICES IN VOCABULARY FORMATION

An important concern in creating the visual vocabulary is the choice of data used to construct it. Generally, researchers report that the most accurate results are obtained when using the same data source to create the vocabulary as is going to be used for the classification or retrieval task. This can be especially noticeable when the application is for specific-level recognition rather than generic categorization. For example, to index the frames from a particular movie, the vocabulary made from a sample of those frames would be most accurate; using a second movie to form the vocabulary should still produce meaningful results, though likely weaker accuracy. When training a recognition system for a particular set of categories, one would typically sample descriptors from training examples covering all categories to try and ensure good coverage. That said, with a large enough pool of features taken from diverse images (admittedly, a vague criterion), it does appear workable to treat the vocabulary as "universal" for any future word assignments.

Furthermore, researchers have developed methods to inject supervision into the vocabulary [Moosmann et al., 2006, Perronnin et al., 2006, Winn et al., 2005], and even to integrate the classifier construction and vocabulary formation processes [Yang et al., 2008]. In this way, one can essentially learn an application-specific vocabulary.

The choice of feature detector or interest operator will also have notable impact on the types of words generated. Factors to consider are (1) the invariance properties required, (2) the type of images to be described, and (3) the computational cost allowable. Using an interest operator (*e.g.*, a DoG detector) yields a sparse set of points that is both compact and repeatable due to the detector's automatic scale selection. For specific-level recognition (*e.g.*, identifying a particular object or landmark building), these points can also provide an adequately distinct description. A common rule of thumb is to use multiple complementary detectors; that is, to combine the outputs from a corner-favoring interest operator with those from a blob-favoring interest operator. See Section 8.1.4 of Chapter 8 for a discussion of visual word representations and choices for category-level recognition.

4.2.4   INVERTED FILE INDEXING

Visual vocabularies offer a simple but effective way to index images efficiently with an *inverted file*. An inverted file index is just like an index in a book, where the keywords are mapped to the page numbers where those words are used. In the visual word case, we have a table that points from the word number to the indices of the database images in which that word occurs. For example, in the cartoon illustration in Figure 4.6, the database is processed and the table is populated with image indices in part (a); in part (b), the words from the new image are used to index into that table, thereby directly retrieving the database images that share its distinctive words.

Retrieval via the inverted file is faster than searching every image, assuming that not all images contain every word. In practice, an image's distribution of words is indeed sparse. Since the index maintains no information about the relative spatial layout of the words per image, typically a spatial verification step is performed on the images retrieved for a given query, as we discuss in detail in the following chapter.

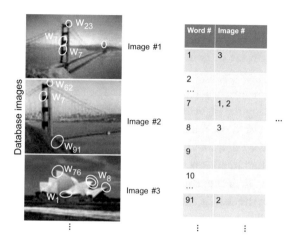

Word #	Image #
1	3
2	
...	
7	1, 2
8	3
9	
10	
...	
91	2

(a) All database images are loaded into the index mapping words to image numbers.

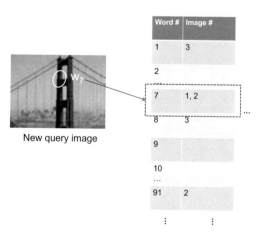

Word #	Image #
1	3
2	
7	1, 2
8	3
9	
10	
...	
91	2

(b) A new query image is mapped to indices of database images that share a word.

Figure 4.6: Main idea of an inverted file index for images represented by visual words.

4.3   CONCLUDING REMARKS

In short, the above methods offer ways to reduce the computational cost of finding similar image descriptors within a large database. While certainly crucial to practical applications of specific object recognition based on local features (the focus of this segment of the lecture), they are also commonly used for other search problems ranging from image retrieval to example-based category recognition, making this section also relevant to generic category algorithms that we will discuss starting in Chapter 7.

Which matching algorithm should be used when? The tree or hashing algorithms directly perform similarity search, offering the algorithm designer the most control on how candidate matches are gathered. In contrast, a visual vocabulary corresponds to a fixed quantization of a vector space, and lacks such control. On the other hand, a visual vocabulary approach has the ability to compactly summarize all local descriptors in an image or window, allowing a fast check for overall agreement between two images. In general, the appropriate choice for an application will depend on the similarity metric that is required for the search, the dimensionality of the data, the available online memory, and the offline resources for data structure setup or other overhead costs.

At this point, we have shown how to detect, describe, and match local features. Good local feature matches between images can alone suggest a specific object has been found; however, to discount spurious matches or to recognize an object from very sparse local features, it is important to also perform a geometric verification stage (see Figure 4.7). Thus, the following chapter closes our discussion of specific object recognition with techniques to verify spatial consistency of the matches.

(a) Matched features alone do not ensure a confident object match.

(b) Candidate matches must next be verified for geometric consistency.

Figure 4.7: The candidate feature matches established using the methods described in this chapter may strongly suggest whether a specific object is present, but are typically verified for geometric consistency. In this example, the good appearance matches found in the top right example can be discarded once we find they do not fit a geometric transformation well, whereas those found in the top left example will check out in terms of both appearance and geometric consistency. Courtesy of Ondrej Chum.

CHAPTER 5

Geometric Verification of Matched Features

Once a set of feature matches has been established, the next step is to verify if those matches also occur in a consistent geometric configuration. By this, we mean that the locations and scales of corresponding features in both images are related by a common geometric transformation. The existence of such a transformation can be motivated by the assumption that we observe the same rigid object in both images. If the entire change in the observed feature configurations is caused by a well-defined change of the observing camera (plus some noise), then we should be able to estimate the underlying transformation given enough correspondences. This is examined in detail in Section 5.1. The influences of noise and outliers cause additional problems. Those are dealt with using robust estimation techniques described in Section 5.2.

5.1   ESTIMATING GEOMETRIC MODELS

We seek to express the allowed change between two corresponding feature configurations by an element of the family of linear transformations. Figure 5.1 shows the different levels of linear geometric transformations that are available to us in the image plane. Starting with the square on the far left, the simplest transformation is a pure translation. Adding also a rotation, we arrive at a *Euclidean* transformation, and adding a scale change we get a *similarity* transformation. At this point, the transformation still preserves angles, parallel lines, and distance ratios. This changes when we move to *affine* transformations, which introduce non-uniform scaling and skew and only preserve parallelism and volume ratios. At the far right end, we see the effect of a *projective* transformation, which only preserves intersection and tangency. A projective transformation of a plane onto another plane is called a *homography*.

Depending on the application and the assumptions we can make about the observed scene, different levels of transformations may be suitable. In the following, we therefore derive methods for estimating *similarity transformations*, *affine transformations*, and *homographies*. In all cases, we make the assumption that all feature correspondences are correct (meaning that we do not have to deal with outliers yet), but may be subject to noise.

For the following derivations, it is often convenient to work in *homogeneous coordinates*, which make it possible to express a translation and projection as a single matrix operation. In homogeneous coordinates, each point vector is extended by an additional coordinate w, $e.g.$, $\mathbf{x} = (x, y, w)$. If $w = 0$,

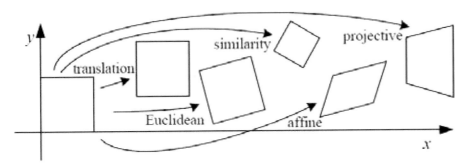

Figure 5.1: The different levels of geometric transformations. Courtesy of Krystian Mikolajczyk.

then the point lies on the *plane at infinity*; else, the point location in ordinary Cartesian coordinates can be obtained by dividing each coordinate by w, *i.e.*, $(x/w, y/w)$.

5.1.1   ESTIMATING SIMILARITY TRANSFORMATIONS

A similarity transformation can already be hypothesized from a single scale- and rotation-invariant interest region observed in both images. Let $f_A = (x_A, y_A, \theta_A, s_A)$ and $f_B = (x_B, y_B, \theta_B, s_B)$ be the two corresponding regions with center coordinates (x, y), rotation θ and scale s. Then we can obtain the transformation from A to B in homogenous coordinates as

$$
T_{\text{sim}} = \begin{bmatrix} ds \cos d\theta & -\sin d\theta & dx \\ \sin d\theta & ds \cos d\theta & dy \\ 0 & 0 & 1 \end{bmatrix}, \text{ where } \begin{aligned} dx &= x_B - x_A \\ dy &= y_B - y_A \\ d\theta &= \theta_B - \theta_A \\ ds &= s_B/s_A \end{aligned}. \tag{5.1}
$$

If only feature locations are available, we require at least two point correspondences. Then we can compute the two vectors between point pairs in the same image, and we obtain T_{sim} as the transformation that projects one such vector onto its corresponding vector in the other image.

5.1.2   ESTIMATING AFFINE TRANSFORMATIONS

Similar to the above, an affine transformation can already be obtained from a single *affine covariant* region correspondence. Recall from Chapter 3 that for estimating the elliptical region shape, we had to compute the region's second-moment matrix M. The transformation that projects the elliptical region onto a circle is given by the square root of this matrix $M^{1/2}$. We can thus obtain the transformation from region $f_A = (x_A, y_A, \theta_A, M_A)$ onto region $f_B = (x_B, y_B, \theta_B, M_B)$ by first projecting f_A onto a circle, rotating it according to $d\theta = \theta_B - \theta_A$, and then projecting the rotated circle back onto f_B. This leads to the following transformation:

$$T_{aff} = M_B^{-1/2} R M_A^{1/2}, \quad \text{with} \quad R = \begin{bmatrix} \cos d\theta & -\sin d\theta & 0 \\ \sin d\theta & \cos d\theta & 0 \\ 0 & 0 & 1 \end{bmatrix}. \tag{5.2}$$

Alternatively, we can estimate the affine transformation from three or more (non-collinear) point correspondences. If more than three such correspondences are available, we can use all of them in order to counteract the influence of noise and obtain a more accurate transformation estimate. This is done as follows. We start by writing down the affine transformation we want to estimate (in non-homogeneous coordinates). This transformation is given by a 2×2 matrix M and a translation vector $\mathbf{t}$, such that

$$\mathbf{x}_B = M\mathbf{x}_A + \mathbf{t}$$
$$\begin{bmatrix} x_B \\ y_B \end{bmatrix} = \begin{bmatrix} m_1 & m_2 \\ m_3 & m_4 \end{bmatrix} \begin{bmatrix} x_A \\ y_A \end{bmatrix} + \begin{bmatrix} t_1 \\ t_2 \end{bmatrix}. \tag{5.3}$$

We can now collect the unknown parameters into one vector $\mathbf{b} = [m_1, m_2, m_3, m_4, t_1, t_2]^\top$ and write the equation in matrix form for a number of point correspondences $\mathbf{x}_{A_i}$ and $\mathbf{x}_{B_i}$:

$$\mathbf{A}\mathbf{b} = \mathbf{X}_B$$
$$\begin{bmatrix} & & \cdots & & & \\ x_{A_i} & y_{A_i} & 0 & 0 & 1 & 0 \\ 0 & 0 & x_{A_i} & y_{A_i} & 0 & 1 \\ & & \cdots & & & \end{bmatrix} \begin{bmatrix} m_1 \\ m_2 \\ m_3 \\ m_4 \\ t_1 \\ t_2 \end{bmatrix} = \begin{bmatrix} \cdots \\ x_{B_i} \\ y_{B_i} \\ \cdots \end{bmatrix}. \tag{5.4}$$

If we have exactly three point correspondences, A will be square, and we can obtain the solution from its inverse as $\mathbf{b} = A^{-1}\mathbf{X}_B$. If more than three correspondences are available, we can solve the equation by building the pseudo-inverse of A:

$$\mathbf{b} = (A^\top A)^{-1} A^\top \mathbf{X}_B = A^\dagger \mathbf{X}_B. \tag{5.5}$$

It can be shown that this solution minimizes the estimation error in the least-squares sense. The results of an affine estimation procedure on a real-world recognition example are shown in Figure 5.2.

5.1.3   HOMOGRAPHY ESTIMATION

A homography, *i.e.*, a projection of a plane onto another plane, can be estimated from at least four point correspondences. When using more than four correspondences, this again has the advantage that we can smooth out noise by searching for a least-squares estimate. Compared to the affine estimation above, the estimation becomes a bit more complicated since we now need to work with

Figure 5.2: Example results of affine transformation estimation for recognition: (top left) model images; (bottom left) test image; (right) estimated affine models and supporting features. From Lowe [1999]. Copyright © 1999 IEEE.

projective geometry. We can do that by using homogeneous coordinates. The homography transformation from a point $\mathbf{x}_A$ to its counterpart $\mathbf{x}_B$ can then be written as follows:

$$\mathbf{x}_B = \frac{1}{z'_B}\mathbf{x}'_B \qquad \text{with} \qquad \mathbf{x}'_B = \mathbf{H}\mathbf{x}_A$$

$$\begin{bmatrix} x_B \\ y_B \\ 1 \end{bmatrix} = \frac{1}{z'_B}\begin{bmatrix} x'_B \\ y'_B \\ z'_B \end{bmatrix} \qquad \begin{bmatrix} x'_B \\ y'_B \\ z'_B \end{bmatrix} = \begin{bmatrix} h_{11} & h_{12} & h_{13} \\ h_{21} & h_{22} & h_{23} \\ h_{31} & h_{32} & 1 \end{bmatrix}\begin{bmatrix} x_A \\ y_A \\ 1 \end{bmatrix} \tag{5.6}$$

The simplest way to estimate a homography from feature correspondences is the *Direct Linear Transformation* (DLT) method [Hartley and Zisserman, 2004]. Using several algebraic manipulations, this method sets up a similar estimation procedure as above, resulting in the following matrix equation for the homography parameters $\mathbf{h}$:

$$\mathbf{A}\mathbf{h} = \mathbf{0}$$

$$\begin{bmatrix} x_{B_1} & y_{B_1} & 1 & 0 & 0 & 0 & -x_{A_1}x_{B_1} & -x_{A_1}y_{B_1} & -x_{A_1} \\ 0 & 0 & 0 & x_{B_1} & y_{B_1} & 1 & -y_{A_1}x_{B_1} & -y_{A_1}y_{B_1} & -y_{A_1} \\ & & & & \cdots & & & & \\ & & & & \cdots & & & & \\ & & & & \cdots & & & & \end{bmatrix}\begin{bmatrix} h_{11} \\ h_{12} \\ h_{13} \\ h_{21} \\ h_{22} \\ h_{23} \\ h_{31} \\ h_{32} \\ 1 \end{bmatrix} = \begin{bmatrix} 0 \\ 0 \\ . \\ . \\ . \end{bmatrix}. \tag{5.7}$$

The solution to this equation is the null-space vector of $\mathbf{A}$. This can be obtained by computing the *singular value decomposition* (SVD) of $\mathbf{A}$, where the solution is given by the singular vector corresponding to the smallest singular value. The SVD of A results in the following decomposition:

$$A = UDV^\top = U \begin{bmatrix} d_{11} & \cdots & 0 \\ \vdots & \ddots & \vdots \\ 0 & \cdots & d_{99} \end{bmatrix} \begin{bmatrix} v_{11} & \cdots & v_{19} \\ \vdots & \ddots & \vdots \\ v_{91} & \cdots & v_{99} \end{bmatrix} \tag{5.8}$$

and the solution for $\mathbf{h}$ is given by the last column of $V^\top$. As above, this solution minimizes the least-squares estimation error.

Since the homography has only 8 degrees of freedom, we are free to bring the result vector into a canonical form by an appropriate normalization. This could be done by normalizing the result vector by its last entry:

$$\mathbf{h} = \frac{[v_{19}, \ldots, v_{99}]}{v_{99}} . \tag{5.9}$$

Although this procedure is often used, it is problematic since v_{99} may also be zero. Hartley and Zisserman [2004] therefore recommend to normalize the vector length instead, which avoids this problem:

$$\mathbf{h} = \frac{[v_{19}, \ldots, v_{99}]}{||[v_{19}, \ldots, v_{99}]||} . \tag{5.10}$$

It should be noted that there are also several more elaborate estimation procedures based on nonlinear optimizations. For those, we however refer the reader to the detailed treatment in [Hartley and Zisserman, 2004].

5.1.4   MORE GENERAL TRANSFORMATIONS

The transformations discussed in this section can also be interpreted in terms of the camera models they afford. *Affine transformations* can only describe the effects of *affine cameras*, a simplified camera model that only allows for orthographic or parallel projection. They are a suitable representation if the effects of perspective distortion are small, such as when the object of interest is far away from the camera and its extent in depth is comparatively small. Affine transformations can also be used to approximate the effects of perspective projection for small regions, such as the local neighborhood of an interest region.

In order to describe the effects of general *perspective cameras*, a *projective transformation* is needed. Section 5.1.3 presented an approach for estimating *homographies*, which capture the perspective projection of a planar surface. In the case of a more general 3D scene, homographies are no longer sufficient, and we need to check if the point correspondences are consistent with an *epipolar geometry*.

If the internal camera calibration is known, the corresponding constraints can be expressed by the so-called *essential matrix*, which captures the rigid 6D transformation (3D translation + 3D rotation) of the camera with respect to a static scene. The essential matrix can be estimated from 5 correspondence pairs. If the internal camera calibration is unknown, we need to estimate the *fundamental matrix*, which can be estimated from 7 correspondence pairs using a non-linear approach or from 8 correspondence pairs using a linear approach. Although those constraints are routinely used in 3D reconstruction, they are, however, only rarely used for object recognition since their estimation is generally less robust. We therefore do not cover them here and refer to [Hartley and Zisserman, 2004] for details.

5.2   DEALING WITH OUTLIERS

The assumption that all feature correspondences are correct rarely holds in practice. In real-world problems, we often have to deal with a large fraction of *outlier correspondences*, *i.e.*, correspondences that are not explained by the chosen transformation model. These outliers can stem from two sources: they can either be caused by wrong or ambiguous feature matches, or they can be due to correct matches that are just not explained by an overly simplistic transformation model (*e.g.*, if an affine transformation model is used to approximate a projective transformation). In both cases, the net effect is the same, namely that the transformed location of a point from one image projected into the other image differs from its correspondence location in that image by more than a certain tolerance threshold.

The problem with outliers is that they can lead to arbitrarily wrong estimation results in connection with *least-squares estimation*. Imagine a simple estimation problem of finding the best-fitting line given a sample of data points. If we use a least-squares error criterion, then moving a single data point sufficiently far away from the correct line will bias the estimated solution towards this point and may move the estimation result arbitrarily far from the desired solution. The same thing will happen with all transformation methods discussed in Section 5.1, since they are also based on least-squares estimation.

In order to obtain robust estimation results, it is therefore necessary to limit the effect of outliers on the obtained solution. For this, we can use the property that the correct solution will result in consistent transformations for all inlier data points, while any incorrect solution will generally only be supported by a smaller, random subset of data points. The recognition task thus boils down to finding a consistent transformation together with a maximal set of inliers supporting this transformation. In the following, we will present two popular approaches for this task: *RANSAC* and the *Generalized Hough Transform*. Both approaches have been successfully used for real-world estimation problems in the past. We will then briefly compare the two estimation schemes and discuss their relative advantages and disadvantages.

5.2.1   RANSAC

RANSAC or *RAndom SAmple Consensus* [Fischler and Bolles, 1981] has become a popular tool for solving geometric estimation problems in datasets containing outliers. RANSAC is a non-deterministic algorithm that operates in a hypothesize-and-test framework. Thus, it only returns a "good" result with a certain probability, but this probability increases with the number of iterations.

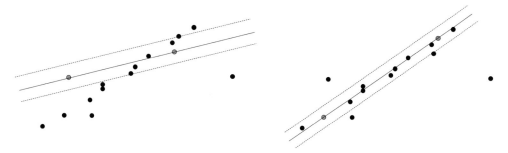

Figure 5.3: Visualization of the RANSAC procedure for a simple problem of fitting lines to a dataset of points in 2D. In each iteration, a minimal set of two points is sampled to define a line, and the number of *inlier points* within a certain distance to this line is taken as its score. In the shown example, the hypothesis on the left has 7 inliers, while the one on the right has 11, making it a better explanation for the observed data. Courtesy of Jinxiang Chai.

Given a set of tentative correspondences, RANSAC randomly samples a minimal subset of m correspondences from this set in order to hypothesize a geometric model (*e.g.*, using any of the techniques described in Section 5.1). This model is then verified against the remaining correspondences, and the number of inliers is determined as its score. This process is iterated until a termination criterion is met. Thus, the RANSAC procedure can be summarized as follows:

1. Sample a minimal subset of m correspondences.

2. Estimate a geometric model T from these m correspondences.

3. Verify the model T against all remaining correspondences and calculate the number of inliers I.

4. If $I > I^\star$, store the new model $T^\star \leftarrow T$, together with its number of inliers $I^\star \leftarrow I$.

5. Repeat until the termination criterion is met (see below).

The RANSAC procedure is visualized in Figure 5.3 for an example of fitting lines to a set of points in the plane. For this kind of problem, the size of the minimal sample set is $m = 2$, *i.e.*, two points are sufficient to define a line in 2D. In each iteration, we thus sample two points to define a line, and we determine the number of inliers to this model by searching for all points within a certain distance to the line. In the example in Figure 5.3, the hypothesis on the left has 7 inliers, while the one on the right has 11 inliers. Thus, the second hypothesis is a better explanation for the observed data and will replace the first one if chosen in the random sampling procedure.

In the above example, RANSAC is applied to the task of finding lines in 2D. Note, however, that RANSAC is not limited to this task, but it can be applied to arbitrary transformation models, including those derived in Section 5.1. In such a case, we define inliers to be those points whose

---

Algorithm 1 RANSAC

$k \leftarrow 0, \varepsilon \leftarrow m/N, I^{\star} \leftarrow 0$
while $\eta = (1 - \varepsilon^m)^k \geq \eta_0$ **do**
    Sample m random correspondences.
    Compute a model T from these samples.
    Compute the number I of inliers for T.
    if $I > I^{\star}$ **then**
        $I^{\star} \leftarrow I, \varepsilon \leftarrow I^{\star}/N$, store T.
    end if
    $k \leftarrow k + 1$.
end while

---

transformation error (*i.e.*, the distance of the transformed point to its corresponding point in the other image), is below a certain threshold.

It can be seen that the more inliers a certain model has, the more likely it is also to be sampled, since any subset of m of its inliers will give rise to a very similar model hypothesis. More generally, an important role in this estimation is played by the true *inlier ratio* $\varepsilon = I^{\star}/N$ of the dataset, *i.e.*, by the ratio of the inliers of the correct solution to all available correspondences. If this ratio is known, then it becomes possible to estimate the number of samples that must be drawn until an uncontaminated sample is found with probability $(1 - \eta_0)$. We can thus derive a run-time bound as follows. Let ε be the fraction of inliers as defined above, and let m be the size of the sampling set. Then the probability that a single sample of m points is correct is ε^m, and the probability that no correct sample is found in k RANSAC iterations is given by

$$\eta = (1 - \varepsilon^m)^k \,. \tag{5.11}$$

We therefore need to choose k high enough, such that η is kept below the desired failure rate η_0. As the true inlier ratio is typically unknown, a common strategy is to use the inlier ratio of the best solution found thus far in order to formulate the termination criterion. The resulting procedure is summarized in Algorithm 1.

RANSAC has proven its worth in a large number of practical applications, in many cases, yielding good solutions already in a moderate number of iterations. As a result of the rather coarse quality criterion (the number of inliers in a certain tolerance band), the initial solution returned by RANSAC will, however, only provide a rough alignment of the model. A common strategy is therefore to refine this solution further, *e.g.*, through a standard least-squares minimization that operates only on the inlier set. However, as such a step may change the status of some inlier or outlier points, an iterative procedure with alternating fitting and inlier/outlier classification steps is advisable.

Since RANSAC's introduction by Fischler & Bolles in 1981, various improvements and extensions have been proposed in the literature

| Model image | Test image | GHT voting space |

Figure 5.4: Visualization of the Generalized Hough Transform (GHT) for object recognition. Each local feature matched between model and test image (shown in yellow) defines a transformation of the entire object reference frame (shown in blue). The GHT lets each such feature pair vote for the parameters of the corresponding transformation and accumulates those votes in a binned voting space. In this example, a 3-dimensional voting space is shown for translation x, y and rotation θ. In practice, scale could be added as a 4th dimension of the voting space. Courtesy of Svetlana Lazebnik, David Lowe, and Lowe [2004], left, and from Leibe, Schindler and Van Gool [2008], right, Copyright © 2008 Springer-Verlag.

(see [*Proc. IEEE Int'l Workshop "25 Years of RANSAC" in conjunction with CVPR*, 2006] for some examples). Among those are extensions to speed up the different RANSAC stages [Capel, 2005, Chum and Matas, 2005, 2008, Matas and Chum, 2004, 2005, Sattler et al., 2009], to deliver run-time guarantees for real-time performance [Nistér, 2003, Raguram et al., 2008], and to improve the quality of the estimated solution [Chum et al., 2004, 2005, Frahm and Pollefeys, 2006, Torr and Zisserman, 2000]. We refer to the rich literature for details.

5.2.2 GENERALIZED HOUGH TRANSFORM

Another robust fitting technique is the *Hough Transform*. The Hough Transform, named after its inventor P.V.C. Hough, was originally introduced in 1962 as an efficient method for finding straight lines in images [Hough, 1962]. Its basic idea is to take the parametric form of a model (*e.g.*, the equation for a line in 2D) and swap the role of the variables and parameters in order to obtain an equivalent representation in the parameter space such that data points lying on the same parametric model are projected onto the same point in parameter space. Ballard [1981] later on showed how this idea could be generalized to detect arbitrary shapes, leading to the *Generalized Hough Transform* (GHT). The basic idea of this extension is that we can let observed single feature correspondences *vote* for the parameters of the transformation that would project the object in the model image to the correct view in the test image.

For this, we use the single-feature estimation approaches described in Sections 5.1.1 and 5.1.2 for scale invariant and affine invariant transformation models. Similar to the above, we subdivide the parameter space into a discrete grid of accumulator cells and enter the vote from each feature correspondence by incrementing the corresponding accumulator cell value. Local maxima in the Hough voting space then correspond to consistent feature configurations and thus to object detection hypotheses. Figure 5.4 visualizes the corresponding GHT procedure for an example of a rotation invariant recognition problem.

As pointed out by Lowe [1999], it is important to avoid all quantization artifacts when performing the GHT. This can be done, *e.g.*, by interpolating the vote contribution into all adjacent cells. Alternatively (or in addition, depending on the level of noise and the granularity of the parameter-space discretization), we can apply Gaussian smoothing on the filled voting space. This becomes all the more important the higher the dimensionality of the voting space gets, as the influence of noise will then spread the votes over a larger number of cells.

5.2.3   DISCUSSION

Comparing RANSAC with the GHT, there is clearly a duality between both approaches. Both try to find a consistent model configuration under a significant fraction of outlier correspondences. The GHT achieves this by starting from a single feature correspondence and casting votes for all model parameters with which this correspondence is consistent. In contrast, RANSAC starts from a minimal subset of correspondences to estimate a model and then counts the number of correspondences that are consistent with this model. Thus, the GHT represents the uncertainty of the estimation in the model parameter space (through the voting space bin size and optional Gaussian smoothing), while RANSAC represents the uncertainty in the image space by setting a bound on the projection error.

The complexity of the GHT is linear in the number of feature correspondences (assuming a single vote is cast for each feature) and in the number of voting space cells. This means that the GHT can be efficiently executed if the size of the voting space is small, but that it can quickly become prohibitive for higher-dimensional data. In practice, a 4D voting space is often considered the upper limit for efficient execution. As a positive point, however, the GHT can handle a larger percentage of outliers with higher dimensionality ($> 95\%$ in some cases), since inconsistent votes are then spread out over a higher-dimensional volume and are thus less likely to create spurious peaks. In addition, the algorithm's runtime is independent of the inlier ratio.

In contrast, RANSAC requires a search through all data points in each iteration in order to find the inliers to the current model hypothesis. Thus, it becomes more expensive for larger datasets and for lower inlier ratios. On the other hand, advantages of RANSAC are that it is a general method suited to a large range of estimation problems, that it is easy to implement, and that it scales better to higher-dimensional models than the GHT. In addition, numerous extensions have been proposed to alleviate RANSAC's shortcomings for a range of problems [*Proc. IEEE Int'l Workshop "25 Years of RANSAC" in conjunction with CVPR*, 2006].

We have now seen the three key steps that state-of-the-art methods use to perform specific object recognition: local feature description, matching, and geometric verification. In the next chapter, we will give examples of specific systems using this general approach.

CHAPTER 6

Example Systems: Specific-Object Recognition

In the following, we will present some applications where the specific object recognition techniques presented above are used in practice. The purpose of this overview is to give the reader a feeling for the range of possibilities, but it should by no means be thought of as an exclusive list.

6.1   IMAGE MATCHING

A central motivation for the development of affine invariant local features was their use for wide-baseline stereo matching. Although this is not directly a recognition problem, it bears many parallels. Thinking of one camera image as the model view, we are interested in finding a consistent set of correspondences in the other view under an epipolar geometry transformation model (not covered in Section 5.1). Figure 6.1 shows an example for such an application in which feature correspondences are first established using affine covariant regions, and RANSAC is then used to find consistent matches [Tuytelaars and Van Gool, 2004].

Figure 6.2 shows another application where local-feature based matching is used for creating panoramas. This approach is again based on SIFT features, which are used to both find overlapping image pairs and estimate a homography between them in order to stitch the images together [Brown and Lowe, 2003, 2007].

6.2   OBJECT RECOGNITION

The introduction of local scale and rotation invariant features such as SIFT [Lowe, 2004] has made it possible to develop robust and efficient approaches for specific object recognition. A popular example is the approach proposed by Lowe [1999, 2004], based on the Generalized Hough Transform described in Section 5.2.2. This approach has been widely used in mobile robotic applications and now forms part of the standard repertoire of vision libraries for robotics.

Figure 6.3 shows recognition results obtained with the GHT in [Lowe, 1999]. The approach described in that paper first extracts scale and rotation invariant SIFT features in each image and then uses matching feature pairs in order to cast votes in a (coarsely-binned) 4-dimensional (x, y, θ, s) voting space. The resulting similarity transformation is in general not sufficient to represent a 3D object's pose in space. However, the Hough voting step provides an efficient way of clustering consistent features by their contribution to the same voting bin. The resulting pose hypotheses

Figure 6.1: Example application: wide-baseline stereo matching. From Tuytelaars and Van Gool [2004], Copyright © 2004 Springer-Verlag.

Figure 6.2: Example application: image stitching. Courtesy of Matthew Brown and from Brown and Lowe [2003, 2007]), Copyright © 2007 Springer-Verlag.

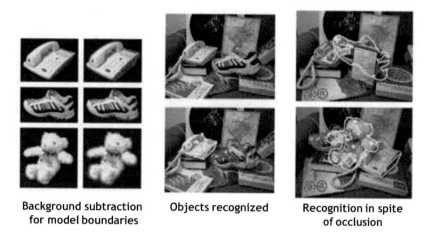

Background subtraction for model boundaries **Objects recognized** **Recognition in spite of occlusion**

Figure 6.3: Object recognition results with the approach by Lowe [1999, 2004] based on the Generalized Hough Transform. Based on Lowe [1999].

Figure 6.4: Example application: large-scale image retrieval. The first column shows a user-specified query region. The other columns contain automatically retrieved matches from a database of about 5,000 images. From Philbin et al. [2007], Copyright © 2007 IEEE.

Figure 6.5: Example application: image auto-annotation. The green bounding boxes show automatically created annotations of interesting buildings in novel test images. Each such bounding box is automatically linked to the corresponding article in Wikipedia. From Gammeter et al. [2009], Copyright © 2009 IEEE.

are then refined by fitting an affine transformation to the feature clusters in the dominant voting bins and counting the number of inlier points as hypothesis score. As a result, the approach can correctly recognize complex 3D objects and estimate their rough pose despite viewpoint changes and considerable partial occlusion.

6.3   LARGE-SCALE IMAGE RETRIEVAL

The techniques from Chapter 3 through 5 make it possible to scale the recognition procedure to very large data sets. A large-scale recognition application making use of this capability was presented by [Philbin et al., 2007] (see Figure 6.4). Here, a number of different affine covariant region detectors are pooled in order to create a feature representation for each database image. The extracted features are stored in an efficient indexing structure (see Chapter 4) in order to allow efficient retrieval from large image databases containing 5,000 to 1,000,000 images. Given a user-specified query region in one image, the system first retrieves a shortlist of database images containing matching features and then performs a geometric verification step using RANSAC with an affine transformation model in order to verify and rank matching regions in other images.

6.4 MOBILE VISUAL SEARCH

A particular field where large-scale content-based image retrieval is actively used today is visual search from mobile phones. Here, the idea is that a user takes a photo of an interesting object from his/her mobile phone and sends it as a query to a recognition server. The server recognizes the depicted object and sends back object-specific content to be displayed on the mobile device.

One of the first approaches to demonstrate practical large-scale mobile visual search was proposed by Nister and Stewenius [2006]. Their approach (based on local features and the Vocabulary Tree indexing scheme described in Section 4.1.1) could recognize examples from a database of 50,000 CD covers in less than a second, while running on a single laptop. In the meantime, a number of commercial services have sprung up that offer mobile visual search capabilities covering databases of several million images, among them *Google goggles* (www.google.com/mobile/goggles/), *kooaba Visual Search* (http://www.kooaba.com/), and *Amazon Remembers*. Almost all such services are based on the local-feature based recognition, matching, and geometric verification pipeline described in the previous chapters. As large databases have to be searched, scalable matching and indexing techniques are key. Despite the large database sizes, the employed techniques have, however, been optimized so far that response times of 1-2s are feasible (including feature extraction, matching, and geometric verification, but without considering communication delays for image transmission). As a result of the local-feature based recognition pipeline, the approaches work particularly well for textured, (locally) planar objects, such as book/CD/DVD covers, movie posters, wine bottle labels, or building facades.

6.5 IMAGE AUTO-ANNOTATION

As a final application example, we present an approach for large-scale image auto-tagging [Gammeter et al., 2009, Quack et al., 2006], *i.e.*, for detection of "interesting" objects in consumer photos and the automatic assignment of meaningful labels or tags to those objects. This is visualized in Figure 6.5. The approach by Gammeter et al. [2009], Quack et al. [2006] starts by automatically mining geotagged photos from internet photo collections and roughly bins them into geospatial grid cells by their geotags. The images in each cell are then matched in order to find clusters of images showing the same buildings (using SURF features and RANSAC with a homography model), which are then also automatically linked to Wikipedia pages through their tags. Given a novel test image, the previously extracted image clusters are used as "beacons" against which the test image is matched (again using SURF features and RANSAC with a homography model). If a match can be established, the matching image region is automatically annotated with the building name, location, and with a link to the associated web content.

6.6 CONCLUDING REMARKS

In this chapter, we have seen how local features can be leveraged for specific object recognition. The general procedure for this task, which was followed by all of the above application examples, was to first match local features between images in order to find candidate correspondences and to then verify their geometric configuration by estimating a common transformation.

We now move on to discuss methods for generic object categorization.

CHAPTER 7

Overview: Recognition of Generic Object Categories

The remainder of this lecture is dedicated to the problem of generic category-level object recognition and detection. Whereas the task of specific object recognition was largely a matching problem, for categories we need to construct models able to cope with the variety of appearance and shape that instances of the same category may exhibit.

The basic steps underlying most current approaches today are as follows:

- First, we must choose a representation and accompanying model (which may be hand-crafted, learned, or some combination thereof).

- Then, given a novel image, we need to search for evidence supporting our object models, and assign scores or confidences to all such candidates.

- Finally, we need to take care to suppress any redundant or conflicting detections.

While the representation and learning choices are vast, we see a natural division between "window-based" models that describe appearance holistically within a region of interest, and "part-based" models that define the appearance of local parts together with some geometric structure connecting them. Thus, in the following, we first describe some commonly used representations, for the window- and part-based genres in turn (Chapter 8).

Given these possible representations, we then discuss how the detection stage proceeds for either model type—from sliding window-based approaches for the holistic representations, to voting and fitting methods for part-based models (Chapter 9). Next, we overview how their associated parameters or discriminative models can be learned from data (Chapter 10). Finally, we close this segment with detailed descriptions of several state-of-the-art and widely used systems that instantiate all of the above components, including the Viola-Jones face detector, HOG person detector, bag-of-words discriminative classifiers, the Implicit Shape Model detector, and the Deformable Part-based Model detector (Chapter 11).

CHAPTER 8

Representations for Object Categories

Many image descriptions can serve generic category models. Our goal in this chapter is to organize and briefly describe a selection of representative examples. At the highest level, we divide the representations into (1) those that are "window-based", meaning that they summarize appearance (texture/shape/geometry...) in a single descriptor for a region of interest (Section 8.1), and (2) those that are "part-based", meaning that they combine separate descriptors for the appearance of a set of local parts together with a geometric layout model (Section 8.2). In terms of appearance, there is significant overlap in the low-level feature used for either type of model; however, we will see in Chapter 9 that their associated procedures for detecting novel instances within a scene differ.

8.1 WINDOW-BASED OBJECT REPRESENTATIONS

How should an image be described to capture the relevant visual cues to recognize an object of a certain category? We briefly overview some candidates in this section, focusing on those types of descriptors that tend to be used to summarize an image window. Note throughout we use "window" interchangeably to refer to the entire image *or* some candidate sub-window within the image. This overview is intended to provide some key pointers of the types of features frequently used in today's recognition systems; however, the list is not exhaustive.

8.1.1 PIXEL INTENSITIES AND COLORS

In order to construct holistic features, we can in principle use the same methods as already introduced in Section 2.1 when we discussed global representations for specific objects. The simplest such representation is a direct concatenation of the pixel intensities into a single feature vector, which can then be optionally processed by subspace methods such as PCA or FLDA. Similarly, we can describe the distribution of colors present with a color histogram; for certain categories, color is a defining cue—the best example, perhaps, being skin [Jones and Rehg, 1999]. On the other hand, in general the importance of color for class discrimination is somewhat limited, both because within many categories there is wide color variation, and since forming illumination-invariant representations of color remains a challenging research issue.

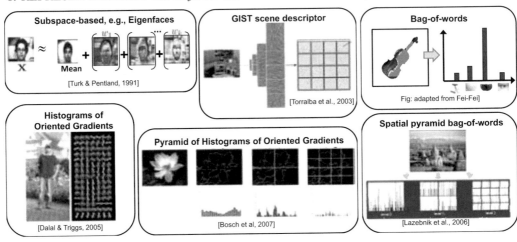

Figure 8.1: Examples of window-based holistic appearance/texture descriptors. From left to right and top to bottom: pixel-based descriptions such as Eigenfaces [Turk and Pentland, 1992] map ordered intensities to a subspace; the GIST descriptor applies multiple steerable filters at multiple scales, and averages responses within a grid of cells in the image [Torralba, 2003]; a bag-of-words descriptor (e.g., [Csurka et al., 2004]) counts the frequency with which each visual word occurs in the image or window; the HOG descriptor histograms the oriented gradients within a grid of overlapping cells [Dalal and Triggs, 2005]; the pyramid of HOG extends this to represent shape within cells of multiple scales [Bosch et al., 2007b]; a spatial bag-of-words histogram computes a bag-of-words histogram within a pyramid of cells over the window [Lazebnik et al., 2006]. Based on Turk and Pentland [1992], Torralba [2003], and Bosch et al. [2007b], and courtesy of Svetlana Lazebnik.

8.1.2   WINDOW DESCRIPTORS: GLOBAL GRADIENTS AND TEXTURE

Aside from raw pixels, features are generally built from the outputs of image filters and other low-level image processing stages. Contrast-based features are of particular interest, due to their insensitivity to lighting changes or color variation. The gradients in an image's intensities indicate edges and texture patterns, the total spatial layout of which comprise an important (detectable) aspect of an object's appearance. There is also evidence that early processing in the human and many animal visual systems relies on gradient or contrast-based representations [Hubel and Wiesel, 1959, 1977]; some work in computer vision takes direct inspiration from biological systems when designing features [Serre et al., 2005].

To reduce sensitivity to small shifts and rotations, effective features often include some form of binning of local image measurements. Spatial histograms offer a way to make the features *locally* orderless, which gives some tolerance to orientation and position differences, while still preserving total layout.

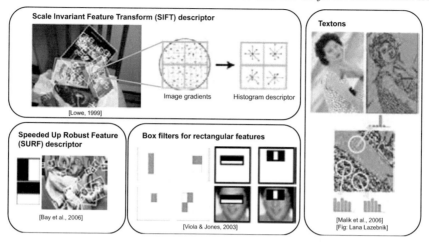

Figure 8.2: Examples of local texture descriptors: SIFT [Lowe, 2004], SURF [Bay et al., 2006], Haar-like box filters [Viola and Jones, 2004], and Textons [Leung and Malik, 1999, Malik et al., 2001]. Based on Lowe [1999], Bay et al. [2006], and Viola and Jones [2004], and courtesy of Svetlana Lazebnik.

Good examples of this concept are the Histogram of Oriented Gradients (HOG) [Dalal and Triggs, 2005], pyramid of HOG [Bosch et al., 2007b], and GIST [Torralba, 2003] descriptors (see Figure 8.1). The HOG descriptor designed by Dalal and Triggs [2005] bins the oriented gradients within overlapping cells. The authors demonstrate its success for pedestrian detection; in general, it is best-suited for textured objects captured from a fairly consistent viewpoint. The pyramid of HOG (pHOG) descriptor [Bosch et al., 2007b] collects the oriented gradient responses in a pyramid of bins pulled from the image or region of interest, summarizing the content within regions of increasingly finer spatial extent. The GIST descriptor of Torralba [2003] is a global representation that divides the image into a 4x4 grid, and within each cell records orientation histograms computed from Gabor or steerable filter outputs. GIST has been shown to be especially valuable as a holistic representation for scene categories.

8.1.3 PATCH DESCRIPTORS: LOCAL GRADIENTS AND TEXTURE

Modifying the general form of the global texture descriptions above to be extracted within only local subwindows or patches, we have a class of local texture/gradient features. See Figure 8.2 for examples.

The SIFT descriptor developed by Lowe [2004] consists of a histogram of oriented image gradients captured within grid cells within a local region. The SIFT descriptor has been shown to be robust under shape distortion and illumination changes, and is therefore widely used. The SURF descriptor [Bay et al., 2006] is an efficient "speeded up" alternative to SIFT that uses simple 2D box filters to approximate derivatives. For details on both descriptors, see Chapter 3.

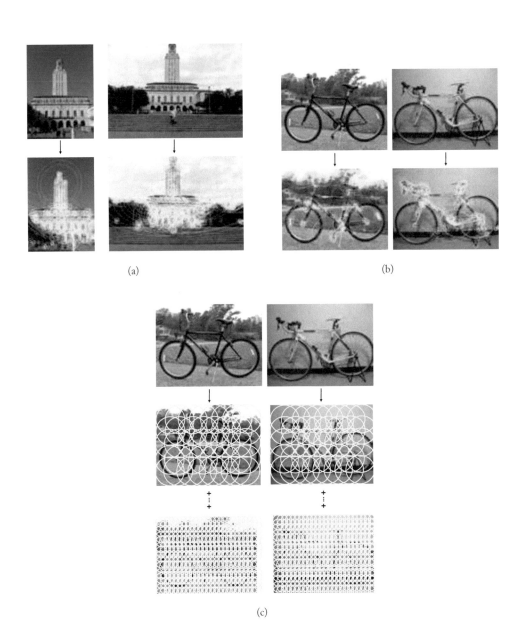

(a)

(b)

(c)

Figure 8.3: Caption on the next page.

Figure 8.3: Depending on the level of recognition and data, *sparse* features detected with a scale invariant interest operator or *dense* multi-scale features extracted everywhere in the image may be more effective. While sparse distinctive points are likely sufficient to match specific instances (like the two images of the UT Tower in (a)), to adequately represent a generic category (like the images of bicycles in (b) and (c)), more coverage may be needed. Note how the interest operator yields a nice set of repeatable detections in (a), whereas the variability between the bicycle images leads to a less consistent set of detections in (b). A dense multi-scale extraction as in (c) will cost more time and memory, but it guarantees more "hits" on the object regions the images have in common. The Harris-Hessian-Laplace detector [Harris and Stephens, 1988] was used to generate the interest points shown in these images.

In the context of specific-object recognition, we considered such patch descriptors in conjunction with a (scale- or affine-) invariant interest operator; that is, we extracted SIFT descriptors at the positions and scales designated by one of the local feature detectors. While for specific objects this is quite effective due to the interest operators' repeatability and descriptors' distinctiveness, for generic category representations, such a sparse set of local features is often insufficient. Instead, for category-level tasks, research suggests that a regular, *dense* sampling of descriptors can provide a better representation [Lazebnik et al., 2006, Nowak et al., 2006]. Essentially, dense features ensure that the object has more regular coverage; there is nothing that makes the "interest" points according to an invariant detector correspond to the semantically interesting parts of an object. When using a dense sampling, it is common to extract patches at a regular grid in the image, and at multiple scales. A compromise on complexity and descriptiveness is to sample randomly from all possible dense multi-scale features in the image. See Figure 8.3.

Viola and Jones define *rectangular features* [Viola and Jones, 2004], which are Haar-like box filters parameterized by their position, scale, and set of internal boxes; for example, the bottom right example on the face in Figure 8.2 signifies that the filter output would be the sum of the intensities in the center white rectangle, minus the sum of the intensities in the left and right black rectangles. While SIFT and SURF are typically extracted at salient interest points detected independently per image, the positions and scales at which rectangular features are extracted is typically learned discriminatively, using a set of labeled images (as in the Viola-Jones frontal face detector; see below in Chapter 11).

Whereas the above local descriptors all pool intensity contrasts within subregions (cells) of the descriptor, *texton* descriptors (e.g., [Leung and Malik, 1999, Malik et al., 2001]) instead record a series of filter bank responses at each pixel, and then summarize the histogram of those responses within local regions. The local summary is computed by mapping each pixel's multidimensional response to a prototypical texton, which are simply representative filter bank responses computed by quantizing a sample of them (i.e., the "visual words" in texture space). The filter bank itself usually consists of convolution kernels sensitive to a variety of scales and orientations. The textons thus give a dense description of the occurrence of basic textures. While original texton-based representations were applied for material classification [Cula and Dana, 2001, Leung and Malik, 1999, Varma and Zisserman, 2002], they are now also often used in object recognition to summarize the texture in a segmented region (e.g., [Shotton et al., 2006]).

8.1.4   A HYBRID REPRESENTATION: BAGS OF VISUAL WORDS

A third window-based representation widely used for generic object category models is the *bag of visual words*. This approach uses a *visual vocabulary*, as introduced earlier in Section 4.2, to compactly summarize the local patch descriptors within a region using a simple 1D histogram (see Figure 4.4 (d)).

It is in some sense a hybrid of the two above mentioned styles (window + local patches), since it records the occurrence of the local visual word descriptors within a window of interest. Whereas the descriptors in Section 8.1.2 (*e.g.*, HOG) record the 2D map of texture with locally orderless component histograms, this representation is completely orderless. This means greater flexibility is allowed (for better or worse) with respect to viewpoint and pose changes. At the same time, the invariance properties of the individual local descriptors make them a powerful way to tolerate viewpoint or pose variation while giving informative local appearance cues.

The regularity or rigidity of an object category's appearance pattern in 2D determines which style is better suited. For example, the class of frontal faces is quite regular and similarly structured across instances, and thus it is more suitable for the 2D layout-preserving descriptors; in contrast, giraffes are articulated, more likely to have pose changes per view, and their coats show inexact variations on a texture pattern—all of which make it suited to a more flexible summary of the texture and key features.

What is particularly convenient about the bag of words (BoW) representation is that it translates a (usually very large) set of high-dimensional local descriptors into a single sparse vector of fixed dimensionality across all images. This in turn allows one to use many machine learning algorithms that by default assume the input space is vectorial—whether for supervised classification, feature selection, or unsupervised image clustering. Csurka et al. [2004] first showed this connection for recognition by using the bag-of-words descriptors for discriminative categorization. Since then, many supervised methods exploit the bag-of-words histogram as a simple but effective representation. In fact, many of the most accurate results in recent object recognition challenges employ this representation in some form [Everingham et al., 2008, Fei-Fei et al., 2004].

Assuming none of the patches in an image overlap, one would get the same description from a BoW no matter where in the image the patches occurred. In practice, features are often extracted such that there is overlap, which at least provides some implicit geometric dependencies among the descriptors. In addition, by incorporating a post-processing spatial verification step, or by expanding the purely local words into neighborhoods and configurations of words (e.g., as in [Agarwal and Triggs, 2006, Lee and Grauman, 2009a, Quack et al., 2007, Savarese et al., 2006]), one can achieve an intermediate representation of the relative geometry. Recent work also considers how to hierarchically aggregate the lowest level tokens from the visual vocabulary into higher level parts, objects, and eventually scenes [Agarwal and Triggs, 2006, Parikh et al., 2009, Sudderth et al., 2005].

To encode global position information into the descriptor, the *spatial pyramid bag of words* computes and concatenates a bag of words histogram in each bin of a pyramid partitioning the image space [Lazebnik et al., 2006] (see Figure 8.1, bottom right). Such a descriptor is sensitive to translations, making it most appropriate for scene-level descriptors, or where the field of view is known to contain the primary object of interest in some regular background.

When the BoW is extracted from the whole image (with spatial binning or otherwise), features arising from the true foreground and those from the background are mixed together, which can be problematic, as the background features "pollute" the object's real appearance. To mitigate this aspect, one can form a single bag from each of an image's segmented regions, or in the case of sliding window classification (to be discussed in Section 9.1), within a candidate bounding box sub-window of the image.

In spite of clear benefits that visual words afford as tools for recognition, the optimal formation of a visual vocabulary remains unclear; building one requires many choices on the part of the algorithm designer. The analogies drawn between textual and visual content only go so far: real words are discrete and human-defined constructs, but the visual world is continuous and yields complex natural images. Real sentences have a one-dimensional structure, while images are 2D projections of the 3D world. Thus, more research is needed to better understand the choices made when constructing vocabularies for local features.

8.1.5   CONTOUR AND SHAPE FEATURES

Whereas the appearance-based features above tend to capture texture and photometric properties, shape-based features emphasize objects' outer boundaries and interior contours. A full discussion on the shape matching problem and shape representations is, however, outside of the scope of this tutorial and we refer to the extensive literature for details [Belongie et al., 2002, Ferrari et al., 2006, Gavrila and Philomin, 1999, Opelt et al., 2006a].

8.1.6   FEATURE SELECTION

Among the many possible descriptors to summarize appearance within an image sub-window, not all necessarily use the entire content of the window. For example, note that in contrast to the HOG descriptor, which uses all pixels within a sub-window, descriptors like the rectangular features (Figure 8.1 center, bottom) are extracted at a specific subset of local regions within the window. Which set to select can be determined by using an interest operator or saliency measure, or else in a discriminative manner.

Discriminative feature selection uses labeled data to identify those feature dimensions or measurements that are most helpful to distinguish between images from different categories. Boosting-based methods can simultaneously select useful features and build a classifier, and they have been employed frequently by recognition methods. Viola and Jones' face detector uses AdaBoost to select

Figure 8.4: While the global object appearance may undergo significant variation inside a category, the appearance and spatial relationship of local parts can often still give important cues. This provides a strong motivation for using part-based models. Courtesy of Rob Fergus and Courtesy of The CalTech256 Dataset.

from a large library of candidate rectangular features—reducing the pool from hundreds of thousands of possible localized filters to the select few (say, hundreds) that are most discriminative between faces and non-faces [Viola and Jones, 2004]. Researchers have also developed methods to jointly select useful features for multiple categories, learning those shared features among classes that yield good discriminative classifiers [Opelt et al., 2006a, Torralba et al., 2004]. Dorko and Schmid [2003] show that classic feature selection measures such as mutual information can be useful to identify discriminative class-specific local features.

8.2   PART-BASED OBJECT REPRESENTATIONS

The previous section introduced object categorization representations based on textured window descriptors or unordered sets of local features (as in the case of bag of words representations). In this section, we examine ways to incorporate more detailed spatial relations into the recognition procedure.

For this, we draw parallels to the specific object recognition techniques presented in Chapters 3 to 6. Back then, we were concerned with establishing exact correspondences between the test image and the model view in order to verify if the matched features occurred in a consistent geometric configuration. As the exact appearance of the model object was known, the extracted features could be very specific, and accurate transformation models could be estimated.

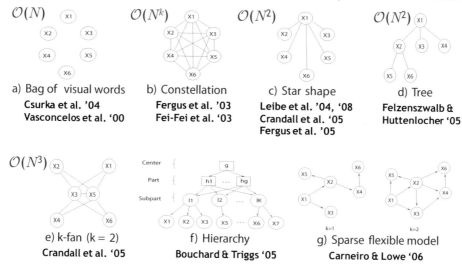

Figure 8.5: Overview of different part-based models investigated in the literature. In this chapter, we focus on two models from this list: the Constellation Model and the Star Model. Based on Carneiro and Lowe [2006].

When moving from specific object recognition to object categorization, however, the task becomes more difficult. Not only may the object appearance change due to intra-category variability, but the spatial layout of category objects may also undergo a certain variation. Thus, we can no longer assume the existence of exact correspondences. However, as shown in Figure 8.4, we can still often find local object fragments or parts with similar appearances that occur in a similar spatial configuration. The basic idea pursued here is therefore to learn object models based on such parts and their spatial relations. The remainder of this section overviews the part-based model structures; later in Chapter 10 we discuss how their parameters are learned from data.

8.2.1   OVERVIEW OF PART-BASED MODELS

Many part-based models have been proposed in the literature. The idea to represent objects as an assembly of parts and flexible spatial relations reaches back to Fischler and Elschlager [1973]. They introduced a model consisting of a number of rigid parts held together by "springs", and formulated a matching objective penalizing both the disagreement between matched parts as well as the springs' tension, thereby constraining the relative movement between parts. While this early work started from a set of hand-defined part templates, most recent approaches try to also learn the part appearance from training data. This implies that the learning algorithm itself should be able to *select* which local object regions to represent, and it should be able to *group* similar local appearances into a common part representation. An optimal solution to the selection problem would imply a search over a huge search space. The development of local invariant features however provides an efficient alternative which has proven to work well in practice. Consequently, most part-based models discussed in the following are based on local features.

Once the parts have been defined, the next question is how to represent their spatial relationship. This choice reflects the mutual independence assumptions we want to make about relative part locations, and it directly affects the number of parameters needed to fully specify the resulting model, as well as the complexity of performing inference using this model.

Various spatial models have been proposed over the years. Figure 8.5 gives an overview of the most popular designs. The simplest model is a bag of visual words, as described in Section 8.1.4 and shown in Fig. 8.5(a). This model does not encode any geometric relations and is listed just for completeness. At the other extreme is a fully connected model, which expresses pairwise relations between any pair of parts. This type of model has become known as a *Constellation Model* and has been used in Fei-Fei et al. [2003], Fergus et al. [2003]. A downside of the full connectivity is that such a model requires an exponentially growing number of parameters as the number of parts increases, which severely restricts its applicability for complex visual categories.

A compromise is to combine the parts in a *Star Model* (Fig. 8.5(c)), where each part is only connected to a central reference part and is independent of all other part locations given an estimate for this reference part. Such a representation has been used in the *Implicit Shape Model* [Leibe et al., 2004, Leibe, Leonardis and Schiele, 2008], in the *Pictorial Structure Model* [Felzenszwalb, Girshick, McAllester and Ramanan, 2010, Felzenszwalb and Huttenlocher, 2005, Felzenszwalb et al., 2008], as well as in several other approaches [Crandall et al., 2005, Fergus, Perona and Zisserman, 2005, Opelt et al., 2006a]. The advantage of this model is its computational efficiency: for N features, exact inference can be performed in $\mathcal{O}(N^2)$ (compared to $\mathcal{O}(N^k)$ for a k-part Constellation model), and more efficient approximations can be devised based on the ideas of the *Generalized Hough Transform* [Leibe et al., 2004, Leibe, Leonardis and Schiele, 2008, Leibe and Schiele, 2003] or the *Generalized Distance Transform* [Felzenszwalb and Huttenlocher, 2005]. The idea of the Star Model can be readily generalized to a *Tree Model* (Fig. 8.5(d)), where each part's location is only dependent on the location of its parent. This type of model is also used in the *Pictorial Structures* framework by Felzenszwalb and Huttenlocher [2005] and has led to efficient algorithms for human pose estimation.

Finally, the above ideas can be generalized in various other directions. The *k-fan Model* [Crandall et al., 2005] (Fig. 8.5(e)) spans a continuum between the fully-connected Constellation Model and the singly-connected Star Model. It consists of a fully-connected set of k reference parts and a larger set of secondary parts that are only connected to the reference parts. Consequently, its computational complexity is in $\mathcal{O}(N^{k+1})$. A similar idea is employed in the *Hierarchical Model* (Fig. 8.5(f)) by Bouchard and Triggs [2005], which contains a (star-shaped) layer of object parts, each of which is densely connected to a set of bottom-level local feature classes. Finally, there is the *Sparse Flexible Model* (Fig. 8.5(g)) proposed by Carneiro and Lowe [2006], where the geometry of each local part depends on the geometry of its k nearest neighbors, allowing for flexible configurations and deformable objects.

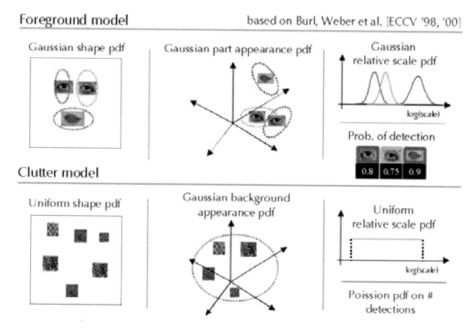

Figure 8.6: Visualization of the different components of the Constellation Model (see text for details). From Fergus et al. [2003]. Copyright © 2003 IEEE.

In the following, we will focus on two models from this list which have been widely used in the literature: the fully connected model and the star-shaped model. We will introduce the basic algorithms behind those approaches and discuss their relative strengths and weaknesses.

8.2.2   FULLY-CONNECTED MODELS: THE CONSTELLATION MODEL

As an example of a fully-connected part representation, we describe the *Constellation Model* by Weber et al. [2000a,b] and Fergus et al. [2003]. This model represents objects by estimating a joint appearance and shape distribution of their parts. Thus, object parts can be characterized either by a distinct appearance or by a distinct location on the object. As a result, the model is very flexible and can even be applied to objects that are only characterized by their texture.

The Constellation model can best be introduced by first considering the recognition task. Given a learned object class model with P parts and parameters θ, the task is to decide whether a new test image contains an instance of the learned object class or not. For this, N local features are extracted with locations $\mathbf{X}$, scales $\mathbf{S}$, and appearances $\mathbf{A}$. The Constellation model now searches for an assignment $\mathbf{h}$ of features to parts in order to make a Bayesian decision R [Fergus et al., 2003]:

$$R = \frac{p(\text{Object}|\mathbf{X}, \mathbf{S}, \mathbf{A})}{p(\text{No object}|\mathbf{X}, \mathbf{S}, \mathbf{A})} \approx \frac{p(\mathbf{X}, \mathbf{S}, \mathbf{A}|\theta)\,p(\text{Object})}{p(\mathbf{X}, \mathbf{S}, \mathbf{A}|\theta_{bg})\,p(\text{No object})}, \tag{8.1}$$

where the likelihood factorizes as follows

$$p(\mathbf{X}, \mathbf{S}, \mathbf{A}|\theta) = \sum_{h \in \mathcal{H}} p(\mathbf{X}, \mathbf{S}, \mathbf{A}, \mathbf{h}|\theta)$$
$$= \sum_{h \in \mathcal{H}} \underbrace{p(\mathbf{A}|\mathbf{X}, \mathbf{S}, \mathbf{h}, \theta)}_{Appearance} \underbrace{p(\mathbf{X}|\mathbf{S}, \mathbf{h}, \theta)}_{Shape} \underbrace{p(\mathbf{S}|\mathbf{h}, \theta)}_{Rel.\ Scale} \underbrace{p(\mathbf{h}|\theta)}_{Other}. \qquad (8.2)$$

That is, we represent the likelihood as a product of separate terms for appearance, shape, relative scale, and other remaining influences. Figure 8.6 shows a visualization of those different components. In each case, a separate model is learned for the object class and for the background.

Briefly summarized, the first term represents each part's appearance independently by a Gaussian density in a 15-dimensional appearance space obtained by PCA dimensionality reduction from 11×11 image patches. This is compared against a single Gaussian density representing the background appearance distribution. The shape term models the joint Gaussian density of the part locations within a hypothesis in a scale-invariant space. The corresponding background clutter model assumes features to be spread uniformly over the image. The scale model is again given by a Gaussian density for each part relative to a common reference frame, also compared against a uniform background distribution. Finally, the last term takes into account both the number of features detected in the image (modeled using a Poisson distribution) and a probability table for all possible occlusion patterns if only a subset of the object parts could be observed.

The classification score is computed by marginalizing over all $|\mathcal{H}| \subseteq \mathcal{O}(N^P)$ possible assignments of features to parts. This marginalization makes it possible to represent an entire category by a relatively small number of parts. It effectively removes the need to make hard assignments at an early stage – if two features provide an equally good support for a certain part, both will contribute substantially to the total classification result. At the same time, the exponential complexity of the marginalization constitutes a major restriction since it limits the approach to a relatively small number of parts.

Figure 8.7 shows the learned representations and recognition results on two different object categories. The first category, motorbikes, has a clearly defined structure. Consequently, the learned model contains well-defined appearances and compact spatial locations for all object parts (as visible from the small covariance ellipses in the upper left plot of the figure). It can also be seen that the parts are consistently found in corresponding locations on the test images, showing that the learned representation really makes sense. In contrast, the second category, "spotted cats" contains significant variability from different body poses and viewing angles. As a result, the Constellation Model focuses on the repeatable texture as the most distinctive feature and keeps only very loose spatial relations. This ability to adapt to the requirements of different categories, automatically weighting the contribution of appearance versus spatial features as needed for the task at hand, is an important property of the Constellation Model.

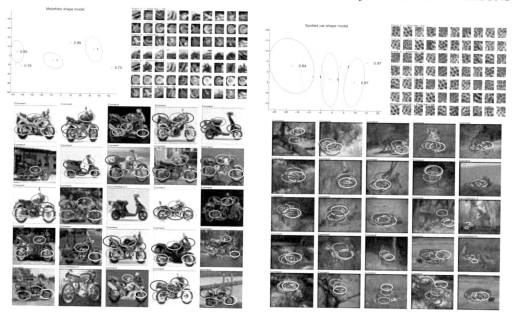

Figure 8.7: Results of the Constellation Model on two object categories: motorbikes and spotted cats. The top row shows the learned representations for spatial relations and appearance; the bottom row contains recognition results on images from the test set, visualizing the best-scoring part assignment. From Fergus et al. [2003]. Copyright © 2003 IEEE.

8.2.3   STAR GRAPH MODELS

The fully-connected shape model described in the previous section is a powerful representation, but suffers from a high computational complexity. In this section, we now examine a recognition approach that builds upon a much simpler spatial representation, namely a Star Model in which each part's location only depends on a central reference part. Given this reference position, each part is treated independently of the others. Thus, the object shape is only defined implicitly by which parts agree on the same reference point. the name of a popular representative of this class of approaches: the *Implicit Shape Model* (ISM) [Leibe et al., 2004, Leibe, Leonardis and Schiele, 2008, Leibe and Schiele, 2003].

The key idea behind recognition in Star Models is to make use of the conditional independence assumption encoded in the model structure in order to efficiently generate hypotheses for the object location. Since each feature is treated independently from all others given the position of the object center, we can learn separate relative location distributions for each of them. If the same feature is then observed on a novel test image, the learned location distribution can be inverted, providing a probability distribution for the object center location given the observed feature location.

In Section 9.2, we will examine two popular approaches employing Star Models for object category recognition, the *Implicit Shape Model* (ISM) by Leibe et al. [2004], Leibe, Leonardis and Schiele [2008] and the *Pictorial Structures Model* by Felzenszwalb and Huttenlocher [2005]. Both employ slightly different strategies for combining the feature contributions and both have been successfully used in practice.

The ISM is motivated by evidence combination in the Generalized Hough Transform. It proposes a probabilistic Hough Voting framework that incorporates the uncertainty inherent in the categorization task. In this framework, the votes from different features are added, and object hypotheses are found as peaks in a (typically 3D $x, y, scale$) voting space. In contrast, the Pictorial Structures Model is motivated by inference in graphical models. It proposes a Generalized Distance Transform that allows fast and exact inference in 2D by integrating the contributions of different parts in a max operation. The *Pictorial Structures* approach can be used both with a Star and with a Tree Model and has become popular both for object detection and for articulated body pose analysis (*e.g.*, in [Andriluka et al., 2008, Ferrari et al., 2008, Ramanan et al., 2007]).

Despite their structural similarities, the two models adopt different philosophical interpretations of what object properties should be represented. The Pictorial Structures model aims at representing a relatively small number of (less than 10) *semantically meaningful parts*, with the tacit assumption that each object of the target category should contain those parts. The parts may undergo appearance variations and may occur in varying spatial configurations, but a majority of them should always be present (and if any part cannot be found, it should be explicitly treated as "occluded"). In contrast, the ISM does not try to model semantically meaningful parts, but instead represents objects as a collection of a large number (potentially 1000s) of *prototypical features* that should ideally provide a dense cover of the object area. Each such prototypical feature has a clearly defined, compact appearance and a spatial probability distribution for the locations in which it can occur relative to the object center. In each test image, only a small fraction of the learned features will typically occur—*e.g.*, different features will be activated for a dark and a brightly colored car—but their consistent configuration can still provide strong evidence for an object's presence.

8.3   MIXED REPRESENTATIONS

Part-based models have the advantage that they can deal with object shapes that are not well-represented by a bounding box with fixed aspect ratio. They have therefore often be applied for recognizing deformable object categories. In contrast, it has been observed that, given sufficient training data, the discriminative window-based models from Section 8.1 (*e.g.*, [Dalal and Triggs, 2005, Viola and Jones, 2004]) often perform better for recognizing mostly rectangular object categories such as faces or front/back views of pedestrians.

Therefore, apart from the clear separation into global and part-based representations, several approaches have been proposed that employ mixtures of the two concepts. The most prominent of those is the *Deformable Part-based Model* by Felzenszwalb et al. [2008]. This model combines a global object template (called a *root filter* in [Felzenszwalb et al., 2008]) matched at a coarse image scale with a star-shaped local part model matched at a finer scale.

The inverse path is taken by Chum and Zisserman [2007]. They start from a Star Model part representation in order to generate initial object hypotheses, which are then verified by comparing the image content to the set of stored exemplar models. Both approaches have achieved good performance in recent PASCAL VOC evaluations [Everingham et al., 2008, PAS, 2007].

8.4   CONCLUDING REMARKS

In this chapter, we have discussed several popular models for object categorization.

The Constellation model was historically one of the first successful part-based models for object categorization. It therefore had a big impact and helped shape the field for the next years. In addition, it initiated a research competition for the best spatial representation and introduced one of the first realistic benchmark datasets for this task. Many of the above-mentioned restrictions were addressed in follow-up work, *e.g.*, in the later papers by [Fei-Fei et al., 2003, Fergus, Perona and Zisserman, 2005]. As research progressed, it became clear that the full connectivity offered by the original Constellation Model was both not required and could not be taken advantage of given the usual training set sizes that were investigated. Instead, star-shaped and tree-shaped spatial models were deemed more promising, as they require far fewer parameters and are more efficient to evaluate. Consequently, Fergus *et al*. themselves proposed an updated version of their model incorporating such a star topology [Fergus, Perona and Zisserman, 2005].

Historically, many common approaches for object detectors have been based on global object models in a sliding-window framework [Dalal and Triggs, 2005, Papageorgiou and Poggio, 2000, Rowley et al., 1998, Viola and Jones, 2004]. This simple design allowed them to benefit directly from advances in classifier development (SVMs, AdaBoost, *etc.*) and powerful discriminative training procedures.

However, in recent years, this situation seems to have changed again, as purely global sliding-window detectors are reaching the performance limits of their representation. At the recent conferences and in the PASCAL VOC'09 evaluation, improved detection performance was reported by several approaches relying again on part-based representations that integrate discriminatively trained part classifiers [Bourdev and Malik, 2009, Felzenszwalb et al., 2008, Gall and Lempitsky, 2009].

CHAPTER 9

Generic Object Detection: Finding and Scoring Candidates

With the basic representations overviewed in the previous chapter, we can now outline the detection process. The detection strategies for window-based and part-based models differ, and so we again discuss them each in turn. Note that the following chapter will describe how the models are learned from image examples.

9.1 DETECTION VIA CLASSIFICATION

In this section, we describe a basic approach that treats category detection as an image classification problem. Assuming we have already decided on a feature representation and trained a classifier that can distinguish the class of interest (say, cars) from anything else using those features, we can then use the decision value of this classifier to determine object presence/absence in a new image. When the object may be embedded amidst "clutter" or background objects, one can insert a *sliding window* search into this pipeline, testing all possible sub-windows of the image with the trained classifier. In that case, the question at each window is, "does this contain object category X or not?" See Figure 9.1 for a sketch.

The sliding window approach truly treats object detection as classification, and asks at every position and scale within the image whether the object is present. To run a multi-scale search, the input image is resampled into a pyramid. The window of interest is then slid through each level, and the classifier outputs are stored. A detector will usually have positive responses at multiple windows nearby the true detection (see Figure 9.2); performing non-maximum suppression as a post-processing step can prune the detections.

How are the detection outputs evaluated? Object detectors can receive partial credit even if the proposed detection bounding box does not agree pixel-for-pixel with the ground truth. Typically, the score used to judge the amount of agreement is the area of intersection of the detection window with the ground truth, normalized by the area of their union. This gives the highest possible score (1.0) for a complete overlap, a score of 0 for a false detection, and an intermediate score when there is some overlap. To make a hard count on correct detections, a threshold is chosen on this overlap score; if the score exceeds the threshold, it is a good detection. Depending on the application, one may prefer to have more false positives or more false negatives. To show the full range of tradeoffs, Receiver Operating Characteristic (ROC) or precision-recall curves are often used, where points on the curve are computed as a function of the threshold on the detector's confidence.

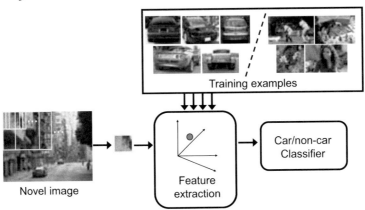

Figure 9.1: Main components of a sliding window detector. To learn from the images, some feature representation must be selected. Labeled examples (positive exemplars, or both negative and positive exemplars) are used to train a classifier that computes how likely it is that a given window contains the object category of interest. Given a novel image, the features from each of its sub-windows at multiple scales are extracted, and then tested by the classifier.

Figure 9.2: Non-maximum suppression is a useful post-processing step to prune out nearby detections.

9.1.1 SPEEDING UP WINDOW-BASED DETECTION

Due to the significant expense of classifying each possible window, techniques have been developed to either prioritize the order in which windows are scanned, or else to quickly eliminate those windows that appear unlikely. Coarse-to-fine detection methods have been developed in which one applies a sequence of tests (or classifiers) to the input window, with each subsequent test designed to be more discriminating, and possibly more expensive, than the last [Amit et al., 2004, Fleuret and Geman, 2001, Rowley et al., 1998, Viola and Jones, 2004]. This approach aims to rule out clear negatives early on, and save more involved computations for those candidate windows that are most likely to belong to the positive class.

For example, Fleuret and Geman [2001] design a chain of tests in this framework; Viola and Jones [2004] propose a cascade structure in which each subsequent detector computes more features, but it is also tuned to produce lower false positive rates. Lampert et al. [2008] devise a branch-and-bound technique for sliding window search that is amenable to certain types of image kernels. By designing an easily computable bound on the classifier output, they show that a priority queue can be used to dramatically reduce the search time in a novel image.

Another strategy is to focus the search for a given object based on localization priors determined by the scene context. Torralba, Murphy, and colleagues show how to prime an object detector according to the global GIST descriptor of the scene [Murphy et al., 2006, Torralba, 2003]. The intuition is that the presence or absence of an object in an image is biased according to what the full scene looks like (*e.g.*, an office scene is more likely to have a computer than a giraffe), and further, the location and scale of a present object depends on the global structures (*e.g.*, a view down a street scene is likely to have small pedestrians in the back, big ones in the front, and an alignment between the detections and the contours of the street). The use of global context is expected to be especially effective when the local appearance in a window is ambiguous, for example, due to low resolution.

9.1.2   LIMITATIONS OF WINDOW-BASED DETECTION

The sliding window approach to detection—and more generally, the treatment of object recognition as an image classification problem—has some important advantages. It is a simple protocol to implement, and offers a ready testbed for sophisticated machine learning algorithms to learn complex patterns. With good features or learning algorithms, this approach to detection can be quite effective. Some of today's most successful methods for faces and pedestrians are in fact built in this framework [Dalal and Triggs, 2005, Viola and Jones, 2001].

On the other hand, there are some clear limitations. First, the high computational cost of searching over all scales and positions is notable. The large number of windows is not only expensive, but it also implies little room for error. With, say, a search over 250,000 locations and 30 orientations and 4 scales, there are 30 million candidate windows in a single image—this puts pressure on having an extremely low false positive rate. Furthermore, treating the detection of each class as its own binary classification problem means that by default, classification time scales linearly with the number of categories learned.

Additionally, the rectangular axis-aligned window used to do scanning is a mismatch for many types of generic categories; not all objects are box-shaped, and so detecting the full extent of the object may require introducing a significant amount of clutter into the features. Similarly, partial occlusions will corrupt the description taken from the window. Since global representations typically assume that all parts of the window should influence the description equally, this is a real problem.

In addition, considering each window independent of the rest of the scene is a handicap, as it entails a loss of context (see Figure 9.3). As such, current research in context-based recognition and joint multi-class object detection aims to move away from the kind of isolated decisions inherent to the sliding window strategy (*e.g.*, [Galleguillos et al., 2008, He et al., 2004, Heitz and Koller, 2008,

Some objects are
almost box-shaped.

Many objects are not.

(a)

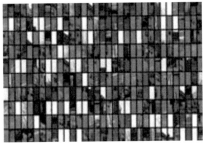

Sliding window

Detector's view

(b)

Figure 9.3: Sliding window detectors have noted limitations. Not all object categories are captured well by a consistent and box-shaped appearance pattern (a), and considering windows in isolation misses out on a great deal of information provided by the scene (b). (a) Left image from Viola and Jones [2001]. Copyright © 2001 IEEE. (b) Courtesy of Derek Hoiem.

Hoiem et al., 2006, Shotton et al., 2006, Singhal et al., 2003, Torralba, 2003]). Such methods make decisions about object presence based on the interaction between objects and scene elements, rather than the appearance of the sub-window alone.

Classification-based methods also assume a sort of regularity in the inputs' spatial layout, so that distances in some feature space are meaningful. For non-rigid, deformable objects or categories with less regular textures, this assumption will not hold. Similarly, this style of detection expects training and testing to be done with the same (roughly) fixed viewpoint relative to the object. For example, to detect profile faces rather than frontal faces, one would need to train a separate classifier.

Finally, training classifiers to recognize the appearance pattern of a category usually entails manually collecting a large number of properly cropped exemplars, which is expensive.

9.2   DETECTION WITH PART-BASED MODELS

The relationship between window-based and part-based approaches can be interpreted as a trade-off between representational complexity and search effort. Window-based approaches consider the entire object appearance holistically. They thus have to achieve the desired level of invariance or generalization to changing object appearance and shape by "blurring" the feature representation or by applying (expensive) non-linear classifiers. In contrast, part-based approaches shift the representational effort to more complex search procedures executed during the detector's run-time.

This strategy makes it easier to deal with partial occlusions and also typically requires fewer training examples. However, its practical success is crucially dependent on efficient methods to search the potentially huge configuration space for consistent constellations of matched parts. In the following, we present several part-based recognition approaches that have been proposed for this step in the literature.

9.2.1   COMBINATION CLASSIFIERS

The simplest approach for combining the information from multiple object parts is by training a combination classifier on their combined output. This approach was taken by Mohan et al. [2001] and Heisele et al. [2001]. Apart from its simplicity, advantages of this approach are that, given enough training data, it can represent complex spatial dependencies between part locations, and that it can account for different reliabilities of the underlying part detectors.

However, combination classifiers also have a number of limitations. By construction, they are restricted to models which consist of a fixed number of parts, and best results are achieved if the part responses themselves are obtained by discriminatively trained detectors [Heisele et al., 2001, Mohan et al., 2001]. In addition, a simple global combination classifier cannot take advantage of possible factorization properties of the part representation. If some object parts can move independently of each other (such as, to some degree, a human's arms and legs), the combination classifier will therefore require more training data in order to represent the entire variability.

9.2.2   VOTING AND THE GENERALIZED HOUGH TRANSFORM

Given a Star Model representation as defined in Section 8.2.3, the different part contributions can be combined in a Generalized Hough Transform (c.f. Section 5.2.2). This idea was popularized by the Implicit Shape Model (ISM), which first proposed an extension of the GHT in order to model the uncertainty inherent in recognizing an object category [Leibe, Leonardis and Schiele, 2008]. See Figure 9.4. Given a new test image, the ISM extracts local features and matches them to the visual vocabulary using soft-matching. Each activated visual word then casts votes for possible positions

  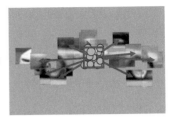

Figure 9.4: Visualization of the basic idea employed in the ISM. (left) During training, we learn the spatial occurrence distribution of each visual word relative to the object center. (middle) For recognition, we use those learned occurrence distributions in order to cast probabilistic votes for the object center in an extension of the Generalized Hough Transform. (right) Once a maximum in the voting space has been found, we can backproject the contributing votes in order to get the hypothesis's support in the image.

of the object center according to its learned spatial distribution, whereupon consistent hypotheses are searched as local maxima in the voting space.

In order to model the uncertainty of the object category, the Hough voting step is formulated with probabilistic weights. The contribution of a feature f observed at location ℓ to the object category o_n at position $\mathbf{x}$ is expressed by a marginalization over all matching visual words $\mathcal{C}_i$:

$$p(o_n, \mathbf{x}|f, \ell) = \sum_i \underbrace{p(o_n, \mathbf{x}|\mathcal{C}_i, \ell)}_{Hough\ vote} \underbrace{p(\mathcal{C}_i|f)}_{Matching\ prob.} . \qquad (9.1)$$

The first term corresponds to the stored occurrence distribution for visual word $\mathcal{C}_i$, which is weighted by the second term, the probability that feature f indeed corresponds to this visual word. In practice, this second term is usually set to $\frac{1}{|\mathcal{C}^\star|}$, where $|\mathcal{C}^\star|$ corresponds to the number of matching visual words. Thus, each image feature casts an entire distribution of weighted votes.

As another difference to the standard GHT, the votes are stored in a continuous 3D voting space for the object position $\mathbf{x} = (x, y, s)$. Maxima in this space are efficiently found using Mean-Shift Mode Estimation [Comaniciu and Meer, 2002] with a scale-adaptive kernel K:

$$\hat{p}(o_n, \mathbf{x}) = \frac{1}{V_b(\mathbf{x}_s)} \sum_k \sum_j p(o_n, \mathbf{x}_j|f_k, \ell_k) K\left(\frac{\mathbf{x} - \mathbf{x}_j}{b(\mathbf{x}_s)}\right), \qquad (9.2)$$

where b is the kernel bandwidth and V_b its volume. Both are adapted according to the scale coordinate $\mathbf{x}_s$, such that the kernel radius always corresponds to a fixed fraction of the hypothesized object size. This way, the recognition procedure is kept scale invariant [Leibe, Leonardis and Schiele, 2008, Leibe and Schiele, 2004].

The search procedure can be interpreted as kernel density estimation for the position of the object center. It should be noted, though, that the ISM voting procedure does not conform to a strict probabilistic model since the vote accumulation implies a summation of probabilities instead of a product combination as would be required. This issue is more closely examined in the recent work by Allan and Williams [2009], Lehmann et al. [2009]. Lehmann et al. [2010] propose a solution motivated by an observed duality to sliding-window detection approaches.

Once a hypothesis has been selected, all features that contributed to it can be backprojected to the image, thereby visualizing the hypothesis's support. As we will show in Section 11.4, this backprojected information can later on be used to infer a top-down segmentation. The main ideas behind the ISM recognition procedure are summarized in Figure 9.4.

9.2.3   RANSAC

As noted in Section 5.2.3, there is a certain duality between the Generalized Hough Transform and RANSAC. Similar to the above-described method based on the GHT, it is therefore also possible to use RANSAC for detection with part-based object category models. This could be advantageous with higher-dimensional transformation spaces (*e.g.*, when not only considering object *location* and *scale*, but also *image-plane rotation*, *aspect ratio*, or other dimensions of variability). Just as in the case of the GHT, it becomes important to represent the variability in the contributing part locations appropriately. For the GHT, this was done by integrating votes in a certain tolerance window (corresponding to the mean-shift kernel in ISM). For a RANSAC based approach, the same role could be fulfilled by the tolerance threshold used for determining inlier features. As for specific object recognition, the main difference is again that the GHT models the estimation uncertainty in the transformation space, while RANSAC represents this uncertainty in the feature location space. Depending on the transformation model and details of the part detection stage, one or the other estimation method may prove superior. Despite those interesting properties, we are however not aware of any current recognition approach making use of RANSAC for part-based object category recognition.

9.2.4   GENERALIZED DISTANCE TRANSFORM

The pictorial structures model goes back to Fischler and Elschlager [1973]. It defines an object as a collection of parts with connections between certain pairs of parts. This can be represented by a graph $G = (V, E)$, where the nodes $V = \{v_1, \ldots, v_n\}$ correspond to the parts and the edges $(v_i, v_j) \in E$ denote the connections. Let $L = \{l_1, ..., l_n\}$ be a certain configuration of part locations. Then the problem of matching the model to an image is formulated as an energy minimization problem using the following energy function:

$$L^* = \arg\min_{L} \left(\sum_{i=1}^{n} m_i(l_i) + \sum_{(v_i, v_j) \in E} d_{ij}(l_i, l_j) \right), \tag{9.3}$$

where $m_i(l_i)$ is the *matching cost* of placing part v_i at location l_i and $d_{ij}(l_i, l_j)$ is the *deformation cost* between two part locations.

Without further restrictions on the model, the optimal part configuration would be quite expensive to compute. However, as Felzenszwalb and Huttenlocher [2005] have shown, if the graph is tree-shaped and if d_{ij} is chosen as a Mahalanobis distance between transformed locations $T_{ij}(l_i)$ and $T_{ji}(l_j)$ with diagonal covariance M_{ij},

$$d_{ij}(l_i, l_j) = (T_{ij}(l_i) - T_{ji}(l_j))^T M_{ij}^{-1} (T_{ij}(l_i) - T_{ji}(l_j)) , \qquad (9.4)$$

then the optimal configuration can be detected in $\mathcal{O}(nh)$, $i.e.$, in time linear in the number of parts n and the number of possible part locations h.

This is done by applying a *Generalized Distance Transform*. The first key idea behind this approach is that, given a tree-shaped graph, the optimal location of any part only depends on its own appearance (via the matching cost) and on the location of its parent (via the deformation cost). Both of those terms can be precomputed. For the matching cost, this is done by independently matching each part with all candidate image locations, resulting in n part response maps of size h. The deformation cost only depends on the relative locations of the part and its parent and can therefore be treated as a fixed template, centered at the part location.

Let us assume for simplicity that the object model has a star shape, $i.e.$, that it corresponds to a tree of depth 2. The matching procedure starts with the part response maps of each leaf node of G. Using the energy minimization formulation from eq. (9.3), the optimal location of the root node can then be expressed as

$$l_1^* = \arg\min_{l_1} \left(m_1(l_1) + \sum_{i=2}^{n} D_{m_i}(T_{1i}(l_1)) \right) \qquad (9.5)$$

$$l_1^* = \arg\min_{l_1} \left(m_1(l_1) + \sum_{i=2}^{n} \min_{l_i} m_i(l_i) + \|l_i - T_{1i}(l_1)\|_{M_{ij}}^2 \right) . \qquad (9.6)$$

That is, we write the cost of each candidate location as a sum of the root node's response map and a distance-transformed version of each child node's response maps. In this view, the response map of each child node is distance transformed with its own specific deformation template.

The second key idea of Felzenszwalb and Huttenlocher [2005] is that when Mahalanobis distances are used as deformation costs, the distance transform can be computed by a separable convolution operation, resulting in the above-mentioned linear-time algorithm.

CHAPTER 10

Learning Generic Object Category Models

This chapter overviews the training procedures and common learning algorithms used to build generic object category models. We first briefly comment on the expected annotations in Section 10.1, and then describe training for window-based (Section 10.2) and part-based (Section 10.3) models in turn.

10.1 DATA ANNOTATION

Current approaches typically rely on supervised learning procedures to build object models, and thus demand some form of manually labeled data prepared by human annotators. This may range from identifying parts of objects, to cropping and aligning images of objects to be learned, to providing complete image segmentations and bounding boxes, to weak image-level tags or other auxiliary data indicating an object's identity. See Figure 10.1 for examples. Typically, the more informative the annotation is, the more manual effort is needed to provide it.

Currently, prominent benchmark datasets provide annotations in one of three forms: (1) "weak supervision", where one is told the name of the primary object of interest within each training image, though it may be surrounded by clutter (see the Caltech-101 and Caltech-256 datasets [Griffin et al., 2007]), (2) bounding box annotations, where any object of interest is outlined with a tight bounding box in the training images (see the PASCAL VOC challenge datasets [Everingham et al., 2008]), and (3) pixel-level labeling, where all objects of interest are segmented with tight polygons in the training images (see the MSRC or LabelMe datasets [MSR-Cambridge, 2005, Russell et al., 2008]). In the latter two styles, typically multiple objects of interest exist in the same image both at training and test time, while the former assumes that test images need only be categorized as a whole.

In practice, the accuracy of most current algorithms improves with larger labeled training sets and more detailed annotations. For example, the very best face detectors are trained with millions of manually labeled examples, and systems built with tedious pixel-level annotations generally outperform those with access only to image-level labels at training time. Given the high manual effort cost of preparing such data, however, researchers are currently exploring ways in which looser forms of supervision, active learning, and semi-supervised learning can be integrated into training object recognition systems (see Section 12.6).

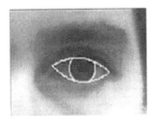

(a) Object parts defined and outlined.

(b) Landmark points marked on object.

(c) Cropped, aligned images of the object.

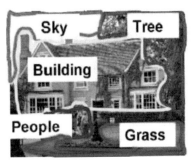

(d) Complete segmentation of multiple objects in scene.

(e) Bounding box on object of interest.

(f) Image-level label specifying the primary object present

Figure 10.1: Caption on the next page.

Figure 10.1: Manual annotation of images used to train recognition systems can take a number of forms—ordered here from approximately most to least manual effort required. (a) from Yuille et al. [1992]; (b) from Cootes et al. [2001]. Copyright © 2001 IEEE; (c) from Viola and Jones [2001]. Copyright © 2001 IEEE; (d) from Vijayanarasimhan and Grauman [2011]. Copyright © 2011 Springer-Verlag; (e) from the PASCAL VOC data set; (f) from the PASCAL VOC image dataset and The Caltech256 dataset, respectively.

10.2   LEARNING WINDOW-BASED MODELS

A wide variety of classification techniques have been used to successfully build detectors operating on window-based representations. One of the first choices one must make is whether to pursue a discriminative or generative model for classification. Discriminative methods directly model the posterior distribution of the class category given the image features, whereas generative methods separately model the class-conditional and prior densities. In effect, this means that a discriminative model focuses the learning effort on the ultimate decision to be made (*e.g.*, is this window showing a car or not), whereas a generative one will further model variability about the category that may be irrelevant to the task. The complexity of these distributions can be difficult to train well with limited labeled data. On the other hand, generative models can also be advantageous since they more often directly support probabilistic estimates, can be used for sampling, allow one to more easily incorporate multiple types of information, and may be interpretable to some degree.

Empirically, discriminative approaches have tended to result in better recognition accuracy in practice. One prepares positive and negative examples, and trains the classifier to distinguish between them. For window-based representations, typically the positive examples consist of tight bounding boxes around instances of the object category, and the negative examples consist of a mix of (1) randomly sampled windows that do not overlap the true instances of that class by more than 50%, and (2) "hard negatives" that an initially trained classifier fires on. For the classifier choice, researchers have explored many techniques such as boosting, nearest neighbor classifiers, support vector machines, neural networks, and conditional random fields. Each has some usual tradeoffs in terms of computational overhead during learning or predictions, amount of free parameters, or sensitivity to outlier training points. Furthermore, usually the particular classifier employed becomes less important the more carefully chosen and engineered the representation is, or the more complete the annotated training set is.

Overall, the classification-based approach to recognition stands in contrast to earlier recognition work in which object models were more "knowledge-rich", perhaps using hand-crafted models to properly capture semantic parts (*e.g.*, the eyes/nose/mouth on the face [Yuille et al., 1989]), or fitting to compositions of human-observed geometric primitives [Biederman, 1987]. Statistical models, on the other hand, leverage labeled training images to directly *learn* what aspects of the appearance are significant for defining the object class.

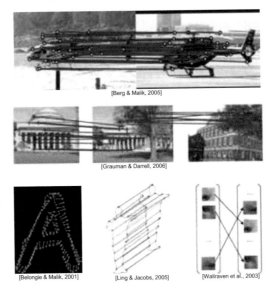

Figure 10.2: Matching functions compute a correspondence between sets of local features. Objects or shapes with a similar set of parts will have a good correspondence between their features, making such measures useful for recognition. Courtesy of Alex Berg, courtesy of Sergie Belongie, and courtesy of Christian Wallraven. Also, from Grauman and Darrell [2006a]. Copyright © 2006 IEEE and from Ling and Jacobs [2007a]. Copyright © 2007 IEEE.

10.2.1 SPECIALIZED SIMILARITY MEASURES AND KERNELS

Both kernel-based learning algorithms (such as support vector machines) and nearest neighbor classifiers are frequently employed in object recognition. Support vector machines (SVMs) can learn patterns effectively in high-dimensional feature spaces and have good generalization performance, while nearest neighbor (NN) classifiers have trivial training requirements and are immediately amenable to multi-class problems. Furthermore, another appealing aspect for both learning algorithms is their modularity with respect to what one chooses for a kernel or similarity function to relate the data instances. Some work in the recognition community has therefore developed kernels specialized for comparing sets of local feature descriptors, which makes it possible to connect the effective SVM or NN classifiers with powerful local representations.

As we saw earlier in the specific-object case, establishing the correspondence between sets of image features is often a telling way to measure their similarity. The one-to-one or many-to-one matches between contour features have long been used to evaluate shape similarity [Belongie et al., 2002, Carlsson, 1998, Chui and Rangarajan, 2000, Gold and Rangarajan, 1996, Johnson and Hebert, 1999, Veltkamp and Hagedoorn, 1999], and most model-based recognition techniques require associating the components of the model template to observed regions in a novel image [Cootes et al., 2001, Felzenszwalb and Huttenlocher, 2005, Weber et al., 2000*b*]. The relatively recent advances in local feature detection and description (reviewed in Chapter 3) have led to an increased focus on matching-based metrics in the recognition community. Generally, the goal is to extract an overall similarity score based on the quality of two images' best correspondence (see Figure 10.2), though methods differ in terms of the particular local descriptors considered and the type of geometric constraints included.

In the following, we first cover the efficient pyramid match kernel approach in detail. Then we describe various ways to preserve geometry in a matching distance, and finally discuss methods based on metric learning that can tailor the distance function to a particular image learning task.

10.2.1.1 The Pyramid Match Kernel

The *pyramid match kernel* (PMK) is a linear-time matching function that approximates the similarity measured by the optimal partial matching between feature sets of variable cardinalities [Grauman and Darrell, 2005, 2007*b*]. Aside from offering an efficient means to judge the similarity between sets of local image features, the PMK also satisfies the Mercer condition for kernel functions. That means that it produces positive-definite Gram matrices and can be used within a number of existing kernel-based machine learning methods, including SVMs.

Consider a feature space $\mathcal{V}$ of d-dimensional vectors for which the values have a maximal range D. The point sets to be matched will come from the input space S, which contains sets of feature vectors drawn from $\mathcal{V}$: $S = \{\mathbf{X}|\mathbf{X} = \{\mathbf{x}_1, \ldots, \mathbf{x}_m\}\}$, where each feature $\mathbf{x}_i \in \mathcal{V} \subseteq \Re^d$, and $m = |\mathbf{X}|$. For example, the vectors could be SIFT descriptors [Lowe, 2004], and a single set would then contain the descriptors resulting from all interest points detected or sampled from a single image.

Given point sets $\mathbf{X}, \mathbf{Y} \in S$, with $|\mathbf{X}| \leq |\mathbf{Y}|$, the optimal partial matching π^* pairs each point in $\mathbf{X}$ to some unique point in $\mathbf{Y}$ such that the total distance between matched points is minimized: $\pi^* = \operatorname{argmin}_\pi \sum_{\mathbf{x}_i \in \mathbf{X}} ||\mathbf{x}_i - \mathbf{y}_{\pi_i}||_1$, where π_i specifies which point $\mathbf{y}_{\pi_i}$ is matched to $\mathbf{x}_i$, and $|| \cdot ||_1$ denotes the L_1 norm. For sets with m features, the Hungarian algorithm computes the optimal match in $O(m^3)$ time [Kuhn, 1955], which severely limits the practicality of large input sizes. In contrast, the pyramid match approximation requires only $O(mL)$ time, where $L = \log D$, and $L \ll m$. In practice, this translates to speedups of several orders of magnitude relative to the optimal match for sets with thousands of features.

The main idea of the pyramid match is to use a multidimensional, multi-resolution histogram pyramid to partition the feature space into increasingly larger regions. At the finest resolution level in the pyramid, the partitions (bins) are very small; at successive levels, they continue to grow in size until a single bin encompasses the entire feature space. At some level along this gradation in bin sizes, any two particular points from two given point sets will begin to share a bin in the pyramid, and when they do, they are considered matched. The key is that the pyramid makes it possible to extract a matching score without computing distances between any of the points in the input sets—the size of the bin that two points share indicates the farthest distance they could be from one another. A

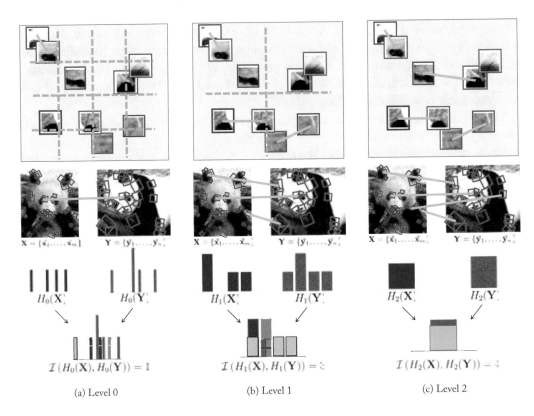

(a) Level 0   (b) Level 1   (c) Level 2

Figure 10.3: Cartoon example of a pyramid match: each set of descriptors is mapped to a multi-resolution histogram in the chosen feature space. Histogram intersection is used to count the number of possible matches at each level of the pyramid, which corresponds to the number of points from the two sets that are within the maximal distance spanned by that histogram level's bin boundaries. Here, there is (a) one possible match at level 0, (b) three possible matches at level 1, and (c) four possible matches at level 2. To score the number of "new" matches, the differences between the successive intersections are computed; here there are $(1 - 0) = 1$, $(3 - 1) = 2$, and $(4 - 3) = 1$ new matches added for (a),(b), and (c), respectively. Each match count is weighted according to the size of the bins at the given level, so that matches between more distant points contribute less significantly to the matching similarity [Grauman and Darrell, 2005].

simple weighted intersection of two pyramids defines an implicit partial correspondence based on the smallest histogram cell where a matched pair of points first appears. See Figure 10.3.

Formally, let a histogram pyramid for input example $\mathbf{X} \in S$ be defined as: $\Psi(\mathbf{X}) = [H_0(\mathbf{X}), \ldots, H_{L-1}(\mathbf{X})]$, where L specifies the number of pyramid levels, and $H_i(\mathbf{X})$ is a histogram vector over points in $\mathbf{X}$. The bins continually increase in size from the finest-level histogram H_0 until the coarsest-level histogram H_{L-1}.[1]

The pyramid match kernel defines similarity between two input sets $\mathbf{Y}$ and $\mathbf{Z}$ as the weighted sum of the number of new matches per level:

$$\mathcal{K}_{PMK}(\Psi(\mathbf{Y}), \Psi(\mathbf{Z})) = \sum_{i=0}^{L-1} w_i \Big(\mathcal{I}(H_i(\mathbf{Y}), H_i(\mathbf{Z})) - \mathcal{I}(H_{i-1}(\mathbf{Y}), H_{i-1}(\mathbf{Z}))\Big),$$

where $\mathcal{I}(H_i(\mathbf{Y}), H_i(\mathbf{Z}))$ denotes the *histogram intersection* between the histograms computed for sets $\mathbf{Y}$ and $\mathbf{Z}$ at pyramid level i, that is, the sum of the minimum value occurring for either of the sets in each bin. This function efficiently extracts the number of new matches at a given quantization level by the difference in successive levels' histogram intersection values. For example, in Figure 10.3, there is one match at the finest scale, two new matches at the medium scale, and one at the coarsest scale. The number of new matches induced at level i is weighted by $w_i = \frac{1}{d2^i}$ to reflect the (worst-case) similarity of points matched at that level. This weighting reflects a geometric bound on the maximal distance between any two points that share a particular bin. Intuitively, similarity between vectors at a finer resolution—where features are more distinct—is rewarded more heavily than similarity between vectors at a coarser level.

As described thus far, the pyramid match cares only about agreement between the appearance of descriptors in two images. That is, the geometry of the features need not be preserved to achieve a high similarity score. One can encode the image position of each feature by concatenating the (x, y) coordinates onto the end of the appearance descriptor, thereby enforcing that matched features must both look the same and occur with a similar layout. However, as is usually the case when adjoining features of different types, this approach may be sensitive to the scaling of the different dimensions.

The *spatial pyramid match kernel*, introduced by Lazebnik and colleagues [Lazebnik et al., 2006], provides an effective way to use the global image positions of features in a pyramid match. The idea is to quantize the appearance feature space into visual words, and then compute the PMK per visual word, where this time the pyramids are built on the space of image coordinates (see Figure 10.4).

Specifically, suppose we have a visual vocabulary comprised of M visual words (see Section 4.2 for a discussion on vocabulary formation). Expand the notation for a set of features $\mathbf{X}$ from above to include each descriptor's image position and word index: $\mathbf{X} = \{(\mathbf{x}_1, x_1, y_1, w_1), \ldots, (\mathbf{x}_m, x_m, y_m, w_m)\}$, where each $w_i \in \{1, \ldots, M\}$. The spatial pyramid is a multi-resolution histogram composed of uniform bins in 2D *image space*. Rather than enter all features into a single pyramid, the spatial pyramid match computes one pyramid per visual word. Thus, each bin of the histogram pyramid for the j-th visual word counts how many times that word occurs

---

[1]For low-dimensional feature spaces, the boundaries of the bins are computed simply with a uniform partitioning along all feature dimensions, with the length of each bin side doubling at each level. For high-dimensional feature spaces (*e.g.*, $d >$ 15), one can use hierarchical clustering to concentrate the bin partitions where feature points tend to cluster for typical point sets [Grauman and Darrell, 2006a], similar to the quantization proposed by Nister *et al.* and discussed in Section 4.2.2. In either case, a sparse representation is maintained per point set that maps each point to its bin indices.

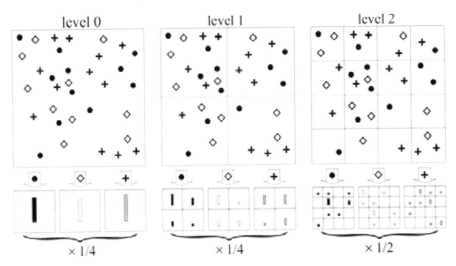

Figure 10.4: The spatial pyramid match kernel matches bags of local features based on both their visual word agreement as well as their closeness in image position. Here the different symbols (plus, circle, *etc.*) denote occurrences of distinct visual words. Matches (intersections) are only counted between words of the same type, and the pyramid bins are constructed in the space of the 2D image coordinates; the total match score sums over all word types. From Lazebnik et al. [2006]. Copyright © 2006 IEEE.

in set $\mathbf{X}$ within its spatial bin boundaries. Denote that pyramid as $\Psi_j(\mathbf{X})$. The spatial pyramid match kernel (SPMK) computes the sum over all words' pyramid match scores:

$$\mathcal{K}_{SPMK}\left(\Psi(\mathbf{Y}), \Psi(\mathbf{Z})\right) = \sum_{j=1}^{M} \mathcal{K}_{PMK}\left(\Psi_j(\mathbf{Y}), \Psi_j(\mathbf{Z})\right). \tag{10.1}$$

With a larger vocabulary, there are more kernel terms to compute and sum; however, since larger vocabularies also increase sparsity, the cost is generally manageable in practice.

The spatial PMK encodes positions of features in a global, translation-sensitive manner; the highest similarity goes to feature sets with the same 2D layout relative to the corners of the image. This kind of spatial regularity is amenable to images of scene categories [Lazebnik et al., 2006] (*e.g.*, coast, mountains, kitchen, dining room), where one can expect the main structures to be laid out consistently within the field of view, or else for images that have been cropped to the object of interest. The 2D geometry constraints are not appropriate for matching unaligned images of objects, although some extensions have shown how to generate regions-of-interest on which to use the spatial PMK [Chum and Zisserman, 2007]. The pyramid match has also been adapted and extended for use within related tasks, such as near-duplicate detection [Xu et al., 2008], medical image classification [Hu et al., 2006], video retrieval [Choi et al., 2008], human action recognition [Lv and Nevatia, 2007], and robot localization [Murilloa and et al., 2007].

10.2.1.2 Matching Distances Accounting for Inter-Feature Geometry

We have already seen a form of spatial *verification* that is often done following feature indexing or voting (as in the Generalized Hough Transform described in Chapter 5), in which specific objects are recognized by verifying that the collected matches together do in fact relate a model (training) image to the input according to a parameterized transformation. In this section, we briefly overview methods that incorporate geometric or spatial relationships among the local features into the set-to-set matching distance itself. For recognition applications, these measures are often used within a simple nearest-neighbor classifier.

One approach is to attempt to compute the best possible assignment between features that minimizes the error between the sets' positions when one is aligned onto the other according to some specified parametric transformation. The error function penalizes for geometric mismatches, and it may also include a term to express the features' appearance agreement as well. For example, the shape context approach of Belongie and colleagues obtains least cost correspondences with an augmenting path algorithm and estimates an aligning thin-plate-spline transform between two input point sets [Belongie et al., 2002]. Such iterative methods have been designed for shape matching [Belongie et al., 2002, Chui and Rangarajan, 2000, Gold and Rangarajan, 1996], where the inputs are assumed to be contour-based image descriptions. Thus, they are best suited for silhouette inputs or sets of edge points (*e.g.*, handwritten digits or trademark symbols). In the special case of closed, solid silhouette shapes, the geometric relationships between points can be treated as a 1D ordering constraint, allowing the use of dynamic programming-style solutions to compute the match cost [Thayananthan et al., 2003].

A linear assignment matching function (as implied for instance by the pyramid match or spatial pyramid match) only penalizes for disagreement between the descriptors themselves, thus capturing only local, "first-order" properties. Recent work shows how recognition can benefit from capturing second-order geometric relationships between the *pairs* of matched features. In contrast to the shape matching methods that fit a parametric transformation, this type of strategy incorporates the constraints between the points in a non-parametric fashion.

For example, Berg and colleagues show how to compute the (approximately) least-cost correspondences between pairs of local features, such that the distortion between matched pairs is also minimal [Berg et al., 2005]. In this case, a matched pair composed of features at points (i, j) in one image and points (i', j') in the other entails a distortion equal to the difference in direction and length of the vectors connecting the pairs in either image. See Figure 10.5(b). In related work, Leordeanu et al. [2007] show how to learn a discriminative set of paired feature parameters on contour features, and use an efficient spectral matching technique to approximate the quadratic assignment problem [Leordeanu and Hebert, 2005]. In either case, since all points participate in paired constraints, the geometry preferences are "interlocked", and computed matches are generally spatially smooth (see image examples in Figure 10.5).

Rather than incorporate the paired constraints into the matching optimization, a more indirect method is to encode the inter-relationships between features as an enhanced input descrip-

(a) Example low-distortion correspondences

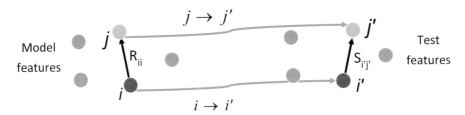

(b) Pairwise distortion between matched features

Figure 10.5: (a) Matching functions that account for the geometry between pairwise local features can yield smooth matching assignments. These example results are from [Berg et al., 2005]. The leftmost images are exemplars with some feature points marked with colored circles. The next image shows the correspondences found, as indicated by the matching colors. White dots are unmatched points. The third image in each row shows all feature points on the exemplar, and the rightmost images show the result of fitting a thin plate spline warp based on the computed correspondences. (b) Sketch of pairwise distortion geometry. If model point j is matched to test point j', and model point i is matched to test point i', we incur a distortion cost relative to the differences in length and direction between the vectors R_{ij} and $S_{i'j'}$. (a) From Berg et al. [2005]. Copyright © 2005 IEEE. (b) Courtesy of Alex Berg.

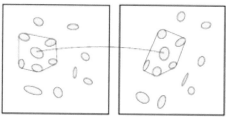

[Sivic & Zisserman, 2004]

[Quack et al. , 2007]   [Lazebnik et al., 2004]

Figure 10.6: Examples of semi-local descriptors that incorporate loose to rigid geometric constraints on top of local features. Left is from Quack et al. [2007], Copyright © 2006 Association for Computing Machinery. Top right is from Sivic and Zisserman [2004], Copyright © 2004, IEEE. Bottom right is from Lazebnik et al. [2004], Copyright © 2004, British Machine Vision Association.

tor [Agarwal and Triggs, 2006, Lazebnik et al., 2004, Lee and Grauman, 2009a, Ling and Soatto, 2007, Quack et al., 2007, Savarese et al., 2006, Sivic and Zisserman, 2004, Yuan et al., 2007]. The general idea is to build more specific features that reflect some geometry and aggregate nearby features into a single descriptor (see Figure 10.6). Discrete visual words are convenient for these kinds of approach, since they allow one to count the co-occurrence and spatial relationships between the features present in an image.

For example, Sivic and Zisserman [2004] propose a loose neighborhood matching definition for configurations of viewpoint invariant features. Let one neighborhood be the convex hull of an elliptical local feature region p and its N spatially nearest neighbors in the same image, and let the second neighborhood be the convex hull of N neighbors surrounding a candidate matching region p' in another image; the configurations are deemed matched if a fraction of the N regions also match (see top right of Figure 10.6). For a neighborhood definition that is more selective about the geometric layout of component features, Lazebnik et al. [2004] propose a correspondence-based approach that identifies groups of local affine regions that remain approximately rigid across multiple instances of an object class (see bottom right of Figure 10.6).

Several other methods capture semi-local neighborhoods that fall in between on the spectrum of geometric rigidity: Savarese et al. [2006] design a *correlaton* feature that encodes how the co-occurrence of a pair of visual words changes as a function of distance, yielding a translation and rotation invariant description of neighborhood appearance layout. Their approach builds on the original definition of correlograms [Huang et al., 1999], which consists of a co-occurrence matrix of pairs of color values as a function of distance. In related work, Ling and Soatto [2007] propose the *proximity distribution kernel* to compare bags of local features while taking into account relative layout. The higher order features are represented by tallying a cumulative distribution of spatially co-occurring visual words, though using the ranked neighborhood ordering rather than absolute distance as above. In work by Agarwal and Triggs [2006], neighborhoods are collected hierarchically in space, while Quack *et al.* propose a tiled region centered on each interest point to bin nearby visual words [Quack et al., 2007] (see left image in Figure 10.6). Lee and Grauman propose a semi-local neighborhood descriptor the reflects both the relative order of spatial proximity as well as the spatial direction in which the features are located with respect to each patch center [Lee and Grauman, 2009a].

Overall, there is a tradeoff between the specificity of the features or matching metrics, versus their flexibility (to viewpoint variation, deformations, *etc.*). Larger and more rigid configurations of local features will be more selective when matching, and higher-order geometric constraints will more strictly preserve layout. On the other hand, with more variable viewing conditions, the weaker feature constraints—and even bags of words models—can offer more robustness. Note that whereas the distances discussed in this section are for matching generic sets of features, below in Section 10.3 we will discuss learning procedures for *part-based models*, which use supervision to select common object parts and then explicitly represent their spatial layout.

10.2.1.3 Learned Metrics for Image-Matching

A standard means of learning about visual categories is to train supervised classifiers from labeled images. Another way to exploit the labeled data is to learn a distance function. A few recent recognition approaches have considered how a learned distance metric can yield better image retrieval or nearest neighbor comparisons. Like the matching functions above, these approaches essentially put more power into the metric, often allowing simpler classifiers to be used on top; however, unlike the methods above, these techniques exploit annotated data to refine the distance function itself.

A good distance metric between images accurately reflects the true underlying relationships, *e.g.*, the category labels or other hidden parameters. It should report small distances for examples that are similar in the parameter space of interest (or that share a class label), and large distances for unrelated examples. Recent advances in metric learning make it possible to learn distance functions that are more effective for a given problem, provided some partially labeled data or constraints are available (see Yang [2006] for a survey). These techniques improve accuracy by taking advantage of the prior information; typically, they optimize parameters to the metric so as to best satisfy the desired constraints.

For example, Frome and colleagues have shown the advantages of learning the weights to attribute to each local image feature when comparing with a correspondence metric [Frome, Singer and Malik, 2007, Frome, Singer, Sha and Malik, 2007]. They introduce a large-margin distance learning approach that aims to preserve relative distance constraints between examples. This serves as a form of discriminative feature selection, identifying those local parts per training example that should have a good match when compared to images from the same category. Jain *et al.* show how to perform scalable similarity search under learned Mahalanobis metrics or kernels, with results improving the accuracy of various base kernels for image comparisons [Jain et al., 2008a].

Another line of work seeks to learn good combinations of kernel functions. Multiple kernel learning takes several kernel matrices as input, together with the associated training labels, and optimizes the weights on each kernel so that the resulting combined kernel is optimally aligned with the "ideal" kernel matrix consisting of maximal values for in-class pairs and minimal values for out-of-class pairs [Bach et al., 2004, Cristianini et al., 2001, Lanckriet et al., 2004]. The promise of multiple kernel learning has been shown by a number of recent results in object recognition and detection [Kumar and Sminchisescu, 2007, la Torre Frade and Vinyals, 2007, Varma and Ray, 2007]. The general approach is useful to automatically choose among kernels parameterized with different values (*e.g.*, the normalization constant in a Gaussian RBF kernel, or the weights associated with each level in the PMK), or to choose among image feature types for a given object categorization task.

Aside from optimizing distances to respect categorical labels (*i.e.*, object class names), metric learning can be employed to tweak a generic measure towards performing better on a visual retrieval task with special preferences. For example, methods have been developed to learn a distance function that will better rank faces according to perceptual similarity [Chopra et al., 2005, Holub et al., 2007], or to select the most effective hash functions mapping similar human 3D pose configurations to the same bucket [Shakhnarovich et al., 2003]. Methods to improve image retrieval by selecting relevant features using boosting have also been explored [Hertz et al., 2004, Tieu and Viola, 2000].

10.3 LEARNING PART-BASED MODELS

Having described methods for learning in window-based models, we now overview the training process for the three representative part-based models presented in the previous chapters.

10.3.1 LEARNING IN THE CONSTELLATION MODEL

The Constellation Model defined in Section 8.2.2 has been designed with the goal of learning with weak supervision. That is, neither the part assignments, nor even object bounding boxes are assumed to be known – only the image labels (target category or background) are provided. Given such a training dataset, the goal of the learning procedure is to find the maximum likelihood estimate for the model parameters $\hat{\theta}_{ML}$, *i.e.*, the parameter setting that maximizes the likelihood for the observed data $\mathbf{X}, \mathbf{S}, \mathbf{A}$ from all training images.

This is achieved using the *expectation maximization* (EM) algorithm. Starting from a random initialization, this algorithm converges to a (locally optimal) solution by alternating between two steps. In the E-step, it computes an expectation for the part assignments given the current value of θ. The M-step then updates θ in order to maximize the likelihood of the current assignment. Since the E-step involves evaluating the likelihood for each of the N^P possible feature-part assignments,

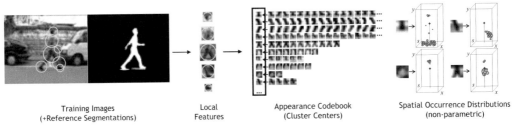

Training Images
(+Reference Segmentations)

Local
Features

Appearance Codebook
(Cluster Centers)

Spatial Occurrence Distributions
(non-parametric)

Figure 10.7: Visualization of the ISM training procedure. From Leibe, Leonardis and Schiele [2008]. Copyright © 2008 Springer-Verlag.

efficient search methods are needed to keep the approach computationally feasible. Still, the authors report training times of 24-36 hours for a single-category model trained on 400 class images in their original paper [Fergus et al., 2003]. This is partially also due to the large number of parameters required to specify the fully-connected model (according to [Fergus et al., 2003], a 5-part model needs 243 parameters and a 6-part model already requires 329 parameters), which in turn impose a requirement on the minimum training set size. Those constraints together restrict the original approach to a small set of only 5-6 parts.

10.3.2 LEARNING IN THE IMPLICIT SHAPE MODEL

The ISM defined in Section 8.2.3 has been designed for supervised learning and requires labeled training examples. In the least case, the labels should include a bounding box for each training object, so that the training algorithm knows the object location and scale. In order to take full advantage of the ISM's subsequent top-down segmentation stage, the training examples should, however, also include a reference segmentation (*e.g.*, given by a polygonal object boundary as available in the LabelMe database [Russell et al., 2008] or by a pixel-wise figure-ground map). Due to its simple Star Model topology, the ISM can treat each object part independently and therefore requires only relatively small training sets (50-150 examples per visual category are usually sufficient).

The full ISM training procedure is visualized in Figure 10.7. The first step is to build up a visual vocabulary (the *appearance codebook*) from scale-invariant local features that overlap with the training objects, using methods as presented in Section 4.2. Next, the ISM learns a *spatial occurrence distribution* for each visual word. For this, we perform a second pass over all training images and match the extracted features to the stored vocabulary using a soft-matching scheme (*i.e.*, activating all visual words within a certain distance threshold). For each visual word, the ISM stores a list of all positions and scales at which this feature could be matched, relative to the object center. This results in a non-parametric probability density representation for the feature position given the object center. As mentioned in Section 8.2.3, the key idea behind the recognition procedure is then that this distribution can be inverted, providing a probability distribution for the object center location given an observation of the corresponding feature.

person   bottle   cat

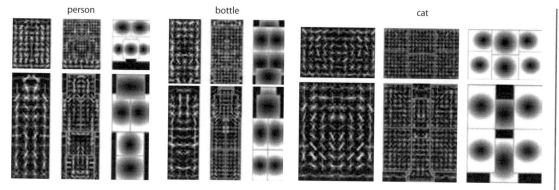

Figure   10.8: Learned   part-based   representations   of   the   Deformable   Part-based
Model [Felzenszwalb, Girshick, McAllester and Ramanan, 2010, Felzenszwalb et al., 2008] for
three categories of the PASCAL dataset [Everingham et al., 2010].
From Felzenszwalb, Girshick, McAllester and Ramanan [2010]. Copyright © 2010 IEEE.

10.3.3 LEARNING IN THE PICTORIAL STRUCTURE MODEL

Part-based models allow the recognition procedure to adapt to deformable object categories and to
deliver more robust recognition results than a holistic detector. As mentioned in Section 8.2.3, a main
difference between the ISM approach and the Pictorial Structures model rests in the interpretation
of what constitutes an object part. While the ISM considers an object as an assembly of potentially
hundreds of local features, the Pictorial Structure model is based on the notion of semantically
meaningful object parts. While this typically restricts the model to a relatively small number of such
parts, an important advantage is that it can reason about the presence or absence of individual parts,
e.g., to encode the knowledge that a human has two legs (and any additional leg found nearby must
belong to a different person).

However, an important challenge for such an approach is to learn the corresponding part
models such that they capture reoccurring and discriminative object structures. While early part-
based approaches relied on hand-labeled part locations [Heisele et al., 2001, Mikolajczyk et al., 2004,
Mohan et al., 2001], such a procedure has the disadvantage of requiring additional manual annotation
and carries the risk of missing effective part locations.

For this reason, Felzenszwalb, Girshick, McAllester and Ramanan [2010], Felzenszwalb et al.
[2008] propose an automatic procedure for learning the part locations from bounding box annotations around the objects of interest. They cast the problem as a *latent SVM* formulation (which is a
special case of a Multiple Instance SVM (MI-SVM)). Given training examples x, part locations z,
resulting feature vectors $\Phi(x, z)$, and model parameters β, the goal is to optimize

$$f_\beta(x) = \max_{z \in Z(x)} \beta \cdot \Phi(x, z) \,, \tag{10.2}$$

where the part locations z for each training example are considered as latent variables. As latent SVMs are non-convex, Felzenszwalb, Girshick, McAllester and Ramanan [2010], Felzenszwalb et al. [2008] resort to an iterative "coordinate descent" training procedure. In each iteration, they

1. *Relabel the positive examples*, *i.e.*, they select the highest scoring part locations z for each positive training example;

2. *Optimize the parameters β* by solving a convex optimization problem.

Note that this procedure does not adapt the latent values for the negative training examples, which would likely not lead to a good model. For efficient training, the optimization step itself is performed using *stochastic gradient descent* (please see [Felzenszwalb, Girshick, McAllester and Ramanan, 2010] for details).

Figure 10.8 shows the models learned with this procedure for several object categories. As can be seen, the learned linear SVM templates are readily interpretable, *e.g.*, showing the human contour and body part locations used in the corresponding detector.

CHAPTER 11

Example Systems: Generic Object Recognition

In the following, we present several successful object recognition approaches in more detail and show how they employ the concepts explained in the previous chapters.

11.1 THE VIOLA-JONES FACE DETECTOR

Frontal faces are a good example of the type of object category well-suited by a window-based representation; there is high regularity in the 2D texture patterns of different face instances if we consider just the rectangular patch centered on the face. The Viola-Jones sliding window face detector is a milestone in generic object detection that uses discriminative learning to discover these regularities [Viola and Jones, 2001]. The technique relies on an ensemble of weak classifiers, each one of which captures some simple contrast-based feature in the candidate face window. While training is fairly expensive and assumes availability of a large number of cropped frontal face instances, detection is very fast due to the use of integral images and an attentional classifier cascade. The detector runs in real time and it is now widely used to find people in images or video.

In the following, we overview the main steps of the Viola-Jones method in order to give a concrete end-to-end instance of training and applying an object detector.

11.1.1 TRAINING PROCESS

- Collect a training set of cropped face images (positive examples) and non-face images (negative examples), all scaled to a common base resolution, e.g., 24×24 pixels.

- Designate a library of candidate rectangular features. Each rectangular feature is a Haar-like filter parameterized by its position relative to the current sliding window location, its scale, and its type (i.e., orientation, number of rectangles). Figure 11.1 shows three examples in the left column. The output of a given feature/filter on an image is the sum of the intensities at pixels within the feature's gray shaded regions, minus the sum of the intensities in the white shaded regions. To give a sense of scale: Viola and Jones designate a bank of 180,000 such features for consideration.

- Compute an *integral image* for each input. The value in an integral image at position (x, y) is the sum of the intensities in the original image above and to the left of (x, y). It is a useful

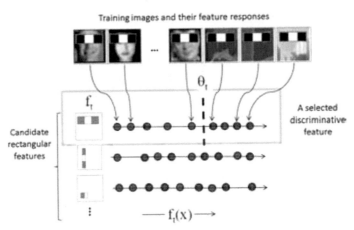

Figure 11.1: The detector training procedure uses AdaBoost to both identify discriminative rectangular features and determine the weak classifiers.

intermediate representation that allows one to compute any rectangular sum within the image using only four array references.

- Compute the rectangular feature responses on all training images. Use the integral image to reduce computational expense; each feature response requires only 6-8 array references.

- Use the AdaBoost algorithm [Freund and Schapire, 1995] and the feature responses computed above to sequentially select a set of discriminative rectangular features. AdaBoost builds a strong classifier out of a set of weighted "weak" classifiers, each of which is performs better than chance according to a weighted error measure on the training data. At each round of boosting, the instances that are incorrectly classified by weak classifiers selected in earlier rounds are given relatively more weight. The intuition is to focus the learners on remaining errors; proofs about the generalization performance justify the strategy formally [Freund and Schapire, 1995, Schapire et al., 1997].

  - For each weak classifier, use a simple decision stump that checks the value of a particular rectangular feature against a threshold. Figure 11.1 illustrates the selection process. Let $f_t(x)$ denote the rectangular feature output for an image x, and let θ_t be the threshold minimizing the weighted error on the training data at round t for that feature. The weak classifier is then

$$h_t(x) = \begin{cases} +1, & \text{if } f_t(x) > \theta_t \\ -1, & \text{otherwise} \end{cases}.$$

(11.1)

  Note that the rectangular feature type and threshold together define one selected feature.

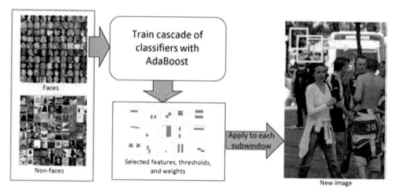

Figure 11.2: Recap of the basic pipeline in the Viola-Jones face detector. Top left face patches are from Viola and Jones [2001].

- The final strong classifier distinguishes face sub-windows from non-face subwindows, and it is the simple weighted sum of the weak classifiers selected with T rounds of boosting:

$$h(\mathbf{x}) = \begin{cases} +1, & \text{if } \sum_{t=1}^{T} \alpha_t h_t(\mathbf{x}) \geq \frac{1}{2} \sum_{t=1}^{T} \alpha_t \\ -1, & \text{otherwise} \end{cases}, \qquad (11.2)$$

where α_t is the classifier weight for round t set by the AdaBoost algorithm, and it is inversely related to the error that particular weak learner had.

11.1.2 RECOGNITION PROCESS

- Given a novel test image, use the sliding window search paradigm: apply the strong classifier to each sub-window in the image in turn. "Each window" entails centering a sub-window at every position and at multiple scales.

- Compute only the selected T rectangular feature responses, again using an integral image for speed.

- Classify the sub-window as face or non-face according to the binary output.

Figure 11.2 summarizes the pipeline outlined thus far. For more details on each step, please see [Viola and Jones, 2001].

11.1.3 DISCUSSION

Typically, detection algorithms are evaluated according to their trade-offs in false positives and detection rates, using ROC or precision-recall curves. Thus, for evaluation against a ground truth

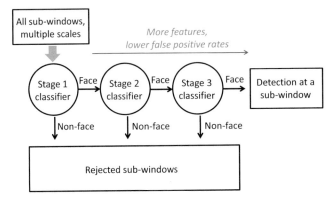

Figure 11.3: Illustration of the attentional cascade. Early classifier stages use few features to remove "obvious" negatives; later stages use more features to discern if the surviving candidate sub-windows may be true detections.

dataset, one would keep the raw classifier outputs ($\sum_{t=1}^{T} \alpha_t h_t(x)$) for all sub-windows and all test images, sort them, and sweep through thresholds from $-\infty$ to ∞ to draw an ROC curve.

Another important aspect of the Viola-Jones detector is its use of a *classifier cascade* to reduce computational time on novel test images (see Section 9.1.1). The attentional cascade they use consists of a series of strong AdaBoost classifiers, where each subsequent classifier stage uses more rectangular features and yields a lower false positive rate. See Figure 11.3. Essentially, classifiers in the earlier stage are cheap to run and immediately dismiss only clear non-face sub-windows. Each sub-window that survives a stage goes on to the next (more expensive) stage, and the process continues. To control the false negative rate in AdaBoost, one reduces the threshold so as to keep a high detection rate with a possibly large false positive rate. For example, in the original Viola-Jones detector, the first stage in the cascade uses only two rectangular features, and it adjusts the threshold to achieve a 100% detection rate and a 40% false positive rate, as determined on a labeled validation set. When building the cascade, one can also isolate helpful "hard negative" instances by taking those non-faces that are false positives according to the cascade thus far.

Note that this face detector is specialized for the viewpoint used in the training images. In other words, we would need to build separate detectors for profile and frontal faces. This is a direct result of the style of features used, which are location-specific relative to the sliding sub-window, and therefore require approximate alignment between instances and a regular 2D texture pattern. For similar reasons, aside from faces, the detector would only be applicable to categories that have a similar structure and shape consistency.

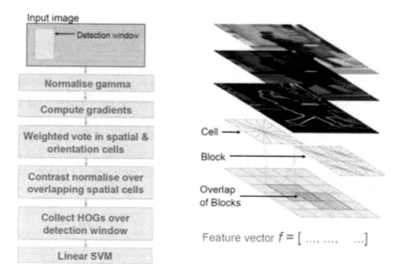

Figure 11.4: Detection procedure for the HOG person detector [Dalal and Triggs, 2005]. Courtesy of Navneet Dalal.

11.2   THE HOG PERSON DETECTOR

As a second example of a discriminative window-based approach object detection, we briefly overview Dalal and Triggs' histogram of oriented gradients (HOG) person detector [Dalal and Triggs, 2005]. As with faces, to some extent upright people ("pedestrians") have a fairly regular 2D texture that is sufficiently captured in a sub-window. In fact, the basic framework shares many aspects with the Viola-Jones face detector described above, with the main distinction being the design of the appearance representation and feature learning.

The main idea of the HOG detector is to extract a dense gradient-based descriptor from the window of interest, and classify it as "person" or "non-person" using a linear support vector machine (SVM) classifier. While closely related descriptors and detection strategies have been explored previously, Freeman and Roth [1995], Lowe [2004], Papageorgiou and Poggio [2000], Dalal and Triggs carefully study the empirical effects of a number of choices in the feature design for person detection—such as the manner of gradient computation, the means of normalization, or grid cell shapes—and show strong performance on benchmark datasets [Dalal and Triggs, 2005].

A HOG descriptor for an image sub-window consists of a concatenated series of gradient histograms. To form the histograms, one first divides the sub-window into small spatial "cells"; each cell produces a histogram based on the gradient values and directions for all pixels within the spatial extent of the cell. (Nine orientation bins and 8×8 pixel cells are used.) As in SIFT and other spatially pooled histograms, this step gives robustness to small shifts or rotations. In order to provide robustness to illumination effects, one contrast-normalizes the local gradient responses before entering them into the histogram. Dalal and Triggs define a block-based normalization that uses a local measure of the energy in regions larger than a cell (see Figure 11.4). The final HOG descriptor for the window consists of the normalized histograms from an overlapping dense grid of blocks.

To train a HOG detector, one simply collects positive and negative examples, extracts their HOG descriptors, and trains a linear SVM in HOG space. To apply the detector on a novel image, one scans the detection window at all positions and scales (using an image pyramid), and then performs non-maximum suppression. See Figure 11.4 for a sketch of the detection process.

In Section 11.5 below, we outline a discriminative deformable part-based model that also uses HOGs and linear filters, but expands the model to include a set of object part filters whose placement can deform relative to the primary ("root") object filter.

11.3   BAG-OF-WORDS IMAGE CLASSIFICATION

As a third example of an end-to-end generic category recognition approach, we overview what is now a standard paradigm for image classification: SVM classification on bag-of-words descriptors. See Figure 11.5. The basic pipeline was first proposed by Csurka et al. [2004], and has since been used and refined by many researchers. Today it is responsible for some of the very best recognition results in the literature; many methods competing on benchmark datasets (e.g., the Caltech or PASCAL VOC) employ some variant of bag-of-words classification.

11.3.1   TRAINING PROCESS

- Collect training image examples from each category of interest.

- Detect or sample local features in all training images. For sparser local features, use one or more interest operators (e.g., DoG, MSER, Harris, etc., see Chapter 3). For denser local features, sample uniformly across the image at multiple scales.

- Extract a descriptor (e.g., SIFT or SURF) at each interest point or sampled point.

- Form a visual vocabulary: sample a corpus of descriptors from the training images, quantize the descriptor space (typically using k-means), and store the resulting "visual words". These words have the same dimensionality as the descriptors (see Section 4.2).

- Map each training image's local descriptors to their respective visual words, yielding a single k-dimensional histogram for each training image.

- Select a kernel function, and use it to compute the pairwise similarities between all training images. For N total training images, we now have an $N \times N$ kernel matrix.

- Use the kernel matrix and labels to train an SVM.

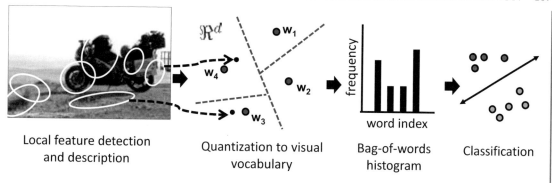

Figure 11.5: Bag-of-words image classification pipeline.

11.3.2 RECOGNITION PROCESS

- Given the novel test image, repeat the feature detection, description, and histogram steps as during training, using the same visual vocabulary.

- Classify the image using the SVM. For non-linear kernels, this entails computing kernel values between the novel bag-of-words histogram and the training instances selected as support vectors.

11.3.3 DISCUSSION

These steps distill the core of a bag-of-words image classification technique. The major choices to be made by the practitioner are which feature extraction method to use (how to sample, which descriptor), how to perform quantization (flat, hierarchical, which clustering algorithm, etc.), and what kernel function to employ. Refer back to Sections 4.2 and 8.1.4 for more background on these choices. The bag-of-words is a holistic representation, and is often used for image-level classification, where we assume that there is a prominent object of interest in the midst of background clutter. However, as usual, it is straightforward to instead train with bounding box annotations and test with a sliding window scan.

Among the most frequently used non-linear kernels are the spatial pyramid match and χ^2-kernel. The χ^2-kernel uses a generalized Gaussian RBF kernel, replacing the usual L_2 norm with the χ^2 distance. The spatial PMK provides coarse spatial pooling for loose layout information. Both kernels are well-suited to histogram representations. We list the SVM classifier specifically due to its widespread use and top empirical results, though of course other classifiers could be interchanged.

Recent research shows that an adaptation of the above feature extraction process to use *sparse coding and max pooling* can further enhance results [Boureau et al., 2010, Wang, Yang, Yu, Lv, Huang and Gong, 2010, Yang et al., 2009]. In that case, one represents each descriptor as a sparse combination of bases (rather than quantizing to a single word) and then takes

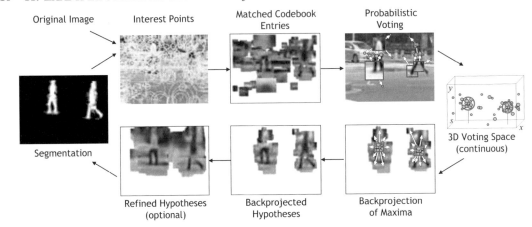

Figure 11.6: Visualization of the ISM recognition procedure. From Leibe, Leonardis and Schiele [2008]. Copyright © 2008 Springer-Verlag.

the maximum within weight associated with each basis in a spatial region (rather than using the total count of each word).

11.4   THE IMPLICIT SHAPE MODEL

The Implicit Shape Model (ISM) provides a successful example for a feature/part-based recognition approach based on a Star Model. In contrast to the three previous approaches, it is not based on discriminatively trained classifiers, but on a relatively simple combination of local feature contributions in a GHT. The ISM and its extensions were successfully applied to a variety of object categories, in particular for detecting pedestrians [Leibe et al., 2005], cars, motorbikes, and cows [Leibe et al., 2007, Leibe, Leonardis and Schiele, 2008]. Compared to window-based approaches such as HOG, a strength of the ISM is that it can deliver robust detection under partial occlusion and significant articulation.

11.4.1   TRAINING PROCESS

The ISM has been designed for supervised learning and requires labeled training examples. In the least case, the labels should include a bounding box for each training object, so that the training algorithm knows the object location and scale. Given such training data, the ISM first extracts and clusters local features in order to create a visual vocabulary and then learns a non-parametric spatial occurrence distribution for each visual word, as described in Section 10.3.2.

11.4.2 RECOGNITION PROCESS

The ISM recognition procedure then follows the idea of the Generalized Hough Transform, as explained in Section 9.2.2. Given a new test image, the ISM extracts local features and matches them to the visual vocabulary using soft-matching. Each activated visual word then casts votes for possible positions of the object center according to its learned spatial distribution, whereupon consistent hypotheses are searched as local maxima in the voting space. This procedure is visualized in Figure 11.6.

11.4.3 VOTE BACKPROJECTION AND TOP-DOWN SEGMENTATION

As already briefly mentioned in Sec. 9.2.2, the ISM recognition procedure does not stop at the voting stage, but also includes a back-projection capability. Once a hypothesis has been selected in the voting space, all features that contributed to it can be back-projected to the image, thereby visualizing the hypothesis's support. The back-projected support already provides a rough indication where the object is in the test image (see Fig. 11.6(bottom)). As the sampled features still contain background structure, this is however not a precise segmentation yet.

The ISM includes a mechanism to refine the back-projected information in order to infer a pixel-wise figure-ground segmentation from the recognition result. For this, it requires a reference segmentation to be provided together with each training example. This reference segmentation could, *e.g.*, be given by a polygonal object boundary (as available in the LabelMe database [Russell et al., 2008]) or by a manually annotated figure-ground map. As the training segmentation only has to be provided for relatively small training sets (50-150 examples per visual category are usually sufficient), this is not an overly big restriction, and the recognition results are significantly improved as a consequence.

The main idea behind the top-down segmentation procedure is that instead of back-projecting the original feature appearance, we can associate any kind of meta-information with the back-projected features. Each vote in the ISM is generated by a matched visual word hypothesizing a particular occurrence location where the corresponding feature was observed on a certain training image. If a reference segmentation is given for this training image, we can thus associate a local figure-ground patch with this feature occurrence. By combining the figure-ground patches associated with all back-projected votes, we obtain a dense figure-ground segmentation, as shown in the last image of Fig. 11.6.

In the following, we present a short derivation of this procedure. We are interested in the probability that a pixel $\mathbf{p}$ is *figure* or *ground* given the object hypothesis, *i.e.*, $p(\mathbf{p}=figure|o_n, \mathbf{x})$. This probability can be obtained by marginalizing over all features containing this pixel and then again marginalizing over all vote contributions from those features to the selected object hypothesis [Leibe, Leonardis and Schiele, 2008, Leibe and Schiele, 2003]:

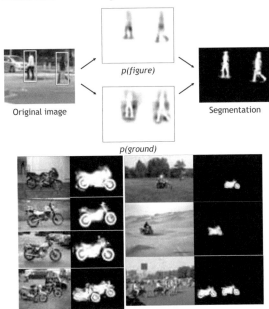

Figure 11.7: (top) Visualization of the ISM top-down segmentation procedure. (bottom) Example detection and segmentation results on motorbikes. From Leibe, Leonardis and Schiele [2008]. Copyright © 2008 IEEE.

$$p(\mathbf{p}=figure|o_n, \mathbf{x}) =$$
$$\sum_{\underbrace{(f_k, \ell_k) \ni \mathbf{p}}_{\substack{\text{all contributing} \\ \text{features containing} \\ \text{pixel } \mathbf{p}}}} \sum_{\underbrace{i}_{\substack{\text{all visual} \\ \text{words matched} \\ \text{to feature } f_k}}} \underbrace{p(\mathbf{p}=fig.|o_n, \mathbf{x}, \mathcal{C}_i, \ell_k)}_{\substack{\text{Stored f/g mask} \\ \text{for each vote}}} \underbrace{p(o_n, \mathbf{x}|f_k, \ell_k)}_{Vote\ weight} \frac{p(f_k, \ell_k)}{p(o_n, \mathbf{x})} \ . \quad (11.3)$$

In this formulation, $p(\mathbf{p}=fig.|o_n, \mathbf{x}, \mathcal{C}_i, \ell_k)$ denotes the stored figure-ground masks for the votes contributed by feature (f_k, ℓ_k). The priors $p(f_k, \ell_k)$ and $p(o_n, \mathbf{x})$ are assumed to be uniform. This means that for every pixel, we effectively build a weighted average over all local segmentation masks stemming from features containing that pixel, where the weights correspond to the features' contribution to the object hypothesis. The final object segmentation is then obtained by computing the likelihood ratio of the *figure* and *ground* probability maps for every pixel, as shown in Figure 11.7.

As shown by Thomas et al. [2007, 2009a], this top-down segmentation procedure can be further generalized to also infer other kinds of meta-data annotations. This includes discrete properties such as part labels, continuous values such as depth maps, as well as vector-valued data such as surface orientations. Some example results are shown in Figure 11.8.

Figure 11.8: As shown by Thomas et al. [2007, 2009a], the ISM top-down segmentation procedure can be generalized to also infer other kinds of meta-data, such as part labels, depth maps, or surface orientations. Based on Thomas et al. [2009a].

11.4.4 HYPOTHESIS VERIFICATION

Finally, the extracted object hypotheses are verified in a model selection procedure, which selects the set of hypotheses that together best explain the image content. Briefly stated, this procedure expresses the score of a hypothesis as the sum over its per-pixel $p(figure)$ probability map. If two hypotheses overlap, then they compete for pixels, as each pixel can only be explained by a single hypothesis. Thus, each pair of hypotheses incurs an interaction cost that is subtracted from their combined scores. A new hypothesis is therefore only selected if it can draw sufficient support from an otherwise as-yet-unexplained image region. This step is important to obtain robust detection performance and significantly improves the recognition results. For details, we refer to Leibe, Leonardis and Schiele [2008].

11.4.5 DISCUSSION

The success of the simple star model representation may first seem surprising, since it imposes no further constraints on relative part locations other than that they should be consistent with a common object center. Clearly, this is quite a weak constraint, but its good performance in practice can

Figure 11.9: (top) Example recognition and segmentation results of the ISM on challenging chair subcategories, demonstrating that given densely sampled features (pooled from three different interest region detectors), the ISM can successfully group also complex wiry shapes. (bottom) An important restriction of the star model used in the ISM is, however, that no higher-level spatial relations between features are encoded. Each local feature that is consistent with the same object center may contribute to a hypothesis. For articulated objects, this may lead to additional body parts being associated to an object hypothesis. From Leibe et al. [2005]. Copyright © 2005 IEEE.

be explained by the large number of local features that contribute to an object hypothesis. If those features overlap, they are no longer truly independent, and consistent responses are enforced this way. This property is also used by the ISM top-down segmentation stage, which further reinforces consistency between overlapping local segmentations. Still, it may happen that additional, spurious object parts are associated to a hypothesis simply because they are also consistent with the same object center. This may particularly become a problem for articulated objects, as shown in Figure 11.9(right). Experience, however, shows that such effects can usually be removed by a further hypothesis verification stage enforcing more global constraints (as done, *e.g.*, in [Leibe et al., 2005]).

The recognition performance, however, hinges on the availability of a sufficient number of input features to cover the target objects. For this reason, later experiments were often based on a combination [Leibe et al., 2006] of several different interest region detectors [Leibe et al., 2007]. For example, the chair detection results shown in Figure 11.9 (left) were obtained through a combination of *Harris-Laplace*, *Hessian-Laplace*, and *DoG* interest regions.

Since its inception, a number of extensions have been proposed for the basic ISM algorithm. Those include adaptations for rotation-invariant voting [Mikolajczyk et al., 2006], multi-cue combination [Leibe et al., 2006, Seemann et al., 2005], multi-category recognition [Leibe et al., 2007, Mikolajczyk et al., 2006], multi-viewpoint detection [Seemann et al., 2006, Thomas et al., 2006], discriminative verification [Fritz et al., 2005, Gall and Lempitsky, 2009, Maji and Malik, 2009], and articulated pose estimation [Andriluka et al., 2008]. We refer to the extensive literature for details.

11.5   DEFORMABLE PART-BASED MODELS

As a final object detection approach, we present the *Deformable Part-based Model* (DPM) proposed by Felzenszwalb, Girshick, McAllester and Ramanan [2010], Felzenszwalb et al. [2008], which has been used with great success on a large number of object categories. It achieved top results for the PASCAL VOC Challenges 2006-2008 [Everingham et al., 2010] and has formed the core of many independent entries in the more recent competitions since then. Thus, this model represents the current state-of-the-art in part-based object detection.

The DPM detector consists of a global *root filter* (similar to the HOG descriptor) and a set of typically 5-6 *part filters*, extracted at a higher resolution. The part locations may vary relative to the root filter location according to a Pictorial Structure [Felzenszwalb and Huttenlocher, 2005] deformation model. Both the part appearances and their location distributions are learned automatically from training data.

11.5.1 TRAINING PROCESS

The DPM detector is trained using the *latent SVM* learning procedure described in Section 10.3.3. Starting from a set of initial bounding box annotations of the category of interest, this procedure automatically learns discriminative part models, while iteratively refining the bounding boxes to best fit the learned model.

11.5.2 RECOGNITION PROCESS

The DPM detector builds upon the HOG representation proposed by Dalal and Triggs [2005]. In its earliest incarnation [Felzenszwalb et al., 2008], it used a slightly simplified form of HOG directly; more recent versions [Felzenszwalb, Girshick, McAllester and Ramanan, 2010] apply a PCA dimensionality reduction on the HOG input in order to speed up computation.

Figure 11.11 shows the DPM recognition pipeline. As said before, the essential idea of the DPM detector is to divide the recognition process into the application of a root filter (similar to the original HOG detector) and of several part filters, extracted at twice the resolution of the

Figure 11.10: DPM recognition results for several categories of the PASCAL dataset [Everingham et al., 2010]. From Felzenszwalb, Girshick, McAllester and Ramanan [2010]. Copyright © 2010 IEEE.

root filter. Each part is associated with a deformation cost that is a (quadratic) function of its displacement with respect to a learned anchor position. The recognition score is then the sum of the filter scores minus the deformation cost. As shown by Felzenszwalb and Huttenlocher [2005], the optimal configuration can be efficiently computed using Generalized Distance Transforms (min-convolutions).

Let $R_{0,l}(x, y)$ be the root filter response and $R_{i,l}(x, y)$ be the response of part filter i applied at position (x, y) and scale level l. Then the detection score can be computed as

$$score(x_0, y_0, l_0) = R_{0,l_0}(x_0, y_0) + \sum_{i=1}^{n} D_{i,l_0-\lambda}(2(x_0, y_0) + v_i) + b , \qquad (11.4)$$

where b is a bias term, v_i is the anchor position of part i, and

$$D_{i,l}(x, y) = \max_{dx,dy} \left\{ R_{i,l}(x + dx, y + dy) - d_i \cdot \phi_d(dx, dy) \right\} \qquad (11.5)$$

is its distance-transformed score using the learned weights d_i of the displacement function $\phi_d(dx, dy)$. After finding a high-scoring root location (x_0, y_0, l_0), the corresponding optimal part locations can then be computed as

$$P_{i,l}(x, y) = \arg \max_{dx,dy} \left\{ R_{i,l}(x + dx, y + dy) - d_i \cdot \phi_d(dx, dy) \right\} . \qquad (11.6)$$

In order to capture different aspects of the target object category, the DPM detector trains several distinct detector models that are combined in a mixture model. During recognition, each component model is applied independently, and their responses are fused in a weighted combination.

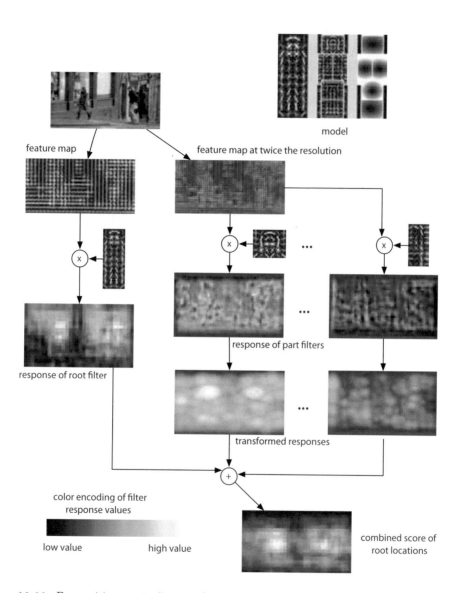

Figure   11.11: Recognition   pipeline   of   the   Deformable   Part-based   Model   detector by   Felzenszwalb, Girshick, McAllester and Ramanan   [2010],   Felzenszwalb et al.   [2008]. From Felzenszwalb, Girshick, McAllester and Ramanan [2010]. Copyright © 2010 IEEE.

In order to further improve performance, the DPM detector predicts a refined object bounding box from the detected part locations. For this, it uses least-squares regression to train a category and aspect specific predictor for each detection model. The resulting detection bounding boxes can optionally be further reranked using context information from other detected object categories. For this, the DPM detector trains a category specific regressor that rescores every detection based on its original score and the highest scoring detection from each other category.

11.5.3   DISCUSSION

Figure 11.10 shows some typical recognition results obtained with the DPM detector for several categories from the PASCAL challenge datasets. As can be seen from those results, the part-based model manages to capture characteristic object structure and leads to good detections despite considerable intra-category variability.

Several recent extensions have been proposed for efficient cascaded evaluation Felzenszwalb, Girshick and McAllester [2010] or for a combination of holistic and part-based detectors in a multi-resolution model Park et al. [2010].

CHAPTER 12

Other Considerations and Current Challenges

Our tutorial thus far has covered the fundamental and (relatively speaking) time-tested techniques in the object recognition literature, emphasizing key representations, matching, learning, and detection strategies to identify visual objects. However, it is certainly not an exhaustive overview of all work ongoing in the research community. In this chapter, we point out other accompanying threads of research on issues central to recognition that are beyond the scope of our lecture, and, for the most part, less mature. We briefly describe the problem area and point the interested reader to recent papers on the topic.

12.1 BENCHMARKS AND DATASETS

Over the last decade, progress in object recognition is quantifiable thanks to the availability of benchmark image datasets with ground truth labels and standard evaluation protocols. These benchmarks provide a common ground for researchers to compare their methods, and they provoke discussion in the community about the types of imagery and annotation on which we should focus. They are frequently used to report results in conference papers. In addition, in recent years, dedicated workshops have been held at major vision meetings for groups to compete with their algorithms on novel withheld test sets. The PASCAL Visual Object Classes Challenge is a prime example [Everingham et al., 2008, 2010].

Recognition datasets vary primarily in terms of the types of imagery (i.e., scenes? object-centric? generic categories? instances? Web images? images captured by the dataset creator?), the types of annotation provided for training (i.e., image-level names, bounding boxes, or pixel-level segmentations), and the test protocol associated with the recognition task (i.e., naming, detection, or pixel labeling). Below we summarize several key datasets and image resources that are currently in use by the recognition community and are available online.

- **Caltech-101 and Caltech-256**: The Caltech datasets [Griffin et al., 2007] consist of images from a wide variety of categories (see Figure 12.1(a) for several example categories). The Caltech-256 (256 categories) is a superset of the Caltech-101 (101 categories). The images are labeled according to which category they exhibit, and there is mostly a single prominent object, with some background clutter. The images originated from keyword search on the Web, and then were pruned by the dataset creators to remove bad examples. The test protocol is a

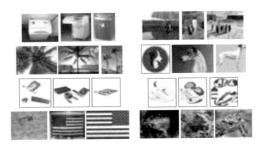

(a) Caltech-256 [Griffin et al., 2007]

(b) PASCAL VOC [Everingham et al., 2008]

(c) LabelMe [Russell et al., 2008]

(d) MSRC [MSR-Cambridge, 2005]

Figure 12.1: Example images and annotations from commonly used datasets.

forced-choice multi-class classification at the image level, and the recognition rate averaged over all 101 or 256 classes is used as the point of comparison.

- **ImageNet**: ImageNet [Deng et al., 2009] is an ongoing collection effort to organize images according to the hierarchy given by nouns in WordNet. There are currently about 11 million total images, with on average 500 images for each of about 15,000 words. Like the Caltech data, images are labeled by name, but there are not further spatial annotations. At the time of writing, ImageNet is not accompanied by a formal benchmark, but an initial taster challenge was given in conjunction with PASCAL in 2010. The Caltech and ImageNet collections offer the largest multi-class recognition challenge in terms of the number of categories. On the other hand, they test image classification only, as opposed to localization or detection.

- **PASCAL VOC**: The PASCAL recognition datasets consist of bounding box labeled images of natural indoor and outdoor scenes, with potentially multiple objects of interest per image. The images originate from Flickr. The associated challenge has been ongoing annually since 2005, and current versions of the dataset have approximately 10,000 images and 20 object categories of interest (person, airplane, cow, sofa, motorcycle, etc; see Figure 12.1(b)). The test protocol is window-based detection, and accuracy is judged according to precision and recall on each category over all images.

- **LabelMe**: LabelMe is an open source annotation tool [Russell et al., 2008]. One can upload images and annotate them through the interface using a polygonal annotation tool (see Figure 12.1(c)). There are more than 11,000 images and 100,000 annotated polygons [Russell et al., 2008]. Since annotators can use free form text to label a polygon (e.g., "car", "automobile", "taxi"), the authors show how to use WordNet to extend and resolve the LabelMe descriptions. There is not a single fixed benchmark associated with LabelMe; users typically select images relevant to the recognition task they wish to examine, and then share the selection for others to make comparisons.

- **MSRC**: The Microsoft Research Cambridge dataset consists of multi-object images of outdoor scenes, labeled by object category at almost every pixel (see Figure 12.1(d)). Version 1 has 240 images and 9 object classes; version 2 has 591 images and 23 object classes. The test protocol is to predict the pixel-level labels on novel images, and so the dataset is frequently used to evaluate recent random field models for image parsing.

We note that the prominence of benchmark datasets in object recognition is a double-edged sword. The sharing of data and annotations has without a doubt helped gauge progress and enables comparisons that would be impossible otherwise. At the same time, the community has to take care not to resist new interesting ideas only because they happen not to "beat" the best numbers available on a given benchmark. Similarly, it is important to distinguish new ideas for recognition from empirical findings that have more to do with overfitting a dataset. Finally, since the choice of annotation style and dataset content themselves bias the task we consider, there is continuing discussion in the field of what choices are most appropriate.

12.2 CONTEXT-BASED RECOGNITION

This tutorial has focused on providing an overview over the current state-of-the-art in specific object identification and object category recognition. However, it is important to consider that objects do not occur in isolation. Rather, they are part of an entire scene and adhere to certain constraints imposed by their environment. For example, many objects rely on a supporting surface, which restricts their possible image locations. In addition, the typical layout of objects in the 3D world and their subsequent projection onto a 2D image leads to a preference for certain image neighborhoods that can be exploited by a recognition algorithm. *Context* therefore plays a crucial role in scene understanding, as has been established in both psychophysical [Biederman, 1981, De Graef et al., 1990, Torralba et al., 2006] and computational [Hoiem et al., 2006, Torralba, 2003] studies.

Different kinds of contextual interactions contribute to this effect. First, there is the *scene context* in which we can expect to observe a certain object. For example, chairs can primarily be seen in indoor settings, whereas cars are a telltale sign of an outdoor environment. This global scene context has, *e.g.*, been used by Torralba et al. [2003] in order to improve object detection performance through contextual priming from scene classification. An important tool for this application are holistic image representations that capture the essential properties of the scene – such as the GIST descriptor proposed by Oliva and Torralba [2001], which includes general spatial properties such as "naturalness", "openness", "roughness", *etc.*

Second, there is *geometric context*, which is a result of the image depicting a real-world scene in which size correlates with distance and occlusion indicates depth ordering. Moreover, many real-world objects rely on a supporting surface, generally a *ground plane*, which restricts their possible locations in the image. Approaches targeted at restricted automotive scenarios have used such geometric constraints for a long time to define ground plane corridors to which object detection can be restricted (*e.g.*, [Bombini et al., 2006, Gavrila and Munder, 2007, Geronimo et al., 2010, Labayrade and Aubert, 2003]). For recognition in more general outdoor settings, Hoiem et al. [2006] have proposed an approach to simultaneously estimate the geometric scene layout and the supporting object detections from a single image. Similar approaches employing stereo image pairs have been proposed by [Leibe et al., 2007, Leibe, Schindler and Van Gool, 2008] and Ess et al. [2007]. More recently, Hedau et al. [2009] have proposed a generalization of Hoiem's approach targeted at indoor images, and Bao et al. [2010] have presented a generalization to multiple supporting planes.

Last but not least, there is *spatial context*, which expresses that certain object or texture regions are more likely to appear next to each other than others. For example, a pedestrian's feet, resting on the ground, are often surrounded by a region containing road surface texture, while his upper body and head are more likely to appear in front of a building facade or surrounded by sky. Spatial context in the form of *neighboring region* information can be modeled with pairwise relations [He et al., 2004], with features that encode inter-pixel or inter-region spatial interactions [Heitz and Koller, 2008, Lee and Grauman, 2010b, Shotton et al., 2006], or by top-down constraints [Singhal et al., 2003]. The benefit of high-level semantic context based on objects' *co-occurrence* information and their *relative locations* has been demonstrated in a number of papers [Desai et al., 2009, Felzenszwalb, Girshick, McAllester and Ramanan, 2010, Galleguillos et al., 2008, Rabinovich et al., 2007], and recent work shows that without such information, impoverished appearance (e.g., due to low resolution) can severely hurt recognition accuracy [Parikh et al., 2008].

Context therefore provides strong cues that should be exploited for recognition. However, this exploitation is often not trivial, as many of the described properties are mutually interdependent. For example, a patch of ground surface often contains few usable cues that could help distinguish it from a wall surface just based on appearance alone. Its most distinguishing feature is that objects are standing on it, but if those have already been detected, how can we still draw benefit from contextual interactions? This indicates that scene interpretation should not be seen as a feed-forward processing chain, but rather as an iterative process.

12.3 MULTI-VIEWPOINT AND MULTI-ASPECT RECOGNITION

Most object category detection approaches discussed so far follow the *appearance-based recognition paradigm* in that they define a visual object category by a set of images showing objects with a common appearance. This is sufficient to build powerful detectors for characteristic viewpoints of many objects, such as rear views of cars or side views of cows. However, it is not sufficient for general recognition of elongated 3D objects, for which the appearance may strongly vary with the viewpoint. In such cases, strategies need to be developed for multi-viewpoint or multi-aspect recognition.

The simplest strategy, which is pursued by many current approaches, is to build up a *battery of single-class detectors* that are independently applied to each image. Such an approach has, *e.g.*, been used by Leibe et al. [2007], Leibe, Schindler and Van Gool [2008] in order to detect multiple viewpoints of cars in street scenes. An improvement on this battery idea is to employ a *mixture model* with separate components for each object aspect, which are fused in a weighted combination, as proposed by Felzenszwalb, Girshick, McAllester and Ramanan [2010].

In the context of the ISM detector, a different resolution strategy has been proposed for multi-aspect detection of pedestrians [Seemann et al., 2006]. This approach first combines features from all different pedestrian viewpoints and articulations in a single detector, but then splits up the contributions according to the different learned aspects as a refinement step. A similar strategy has recently been proposed by Razavi et al. [2010] for multi-aspect car detection with a dense Hough Forest detector [Gall and Lempitsky, 2009].

Methods for specific object recognition can directly encode 3D geometric constraints [Rothganger et al., 2003], which provides strong cues for object detection under viewpoint changes. A string of approaches has tried to make use of similar constraints for object category detection. However, this is far more challenging, since there are generally no direct correspondences between different objects of the same category. Thomas et al. [2006, 2009b] propose a multi-viewpoint extension of the ISM detector that uses automatically estimated geometric constraints between different viewpoints of the same training object in order to learn how spatial relations of learned codebook entries transfer between viewpoints. Hoiem et al. [2007] explicitly encode the dense layout of local patch appearances by mapping them onto a stylized 3D model for multi-view car recognition.

Savarese and Fei-Fei [2007] propose a more general solution for a part-based multi-viewpoint detector that represents general 3D objects by a configuration of planar parts whose relative spatial arrangement is given by pairwise homographies. This approach has the advantage that it can readily adapt to viewpoint changes, while keeping a relatively simple representation. Several more recent papers extend the approach towards an increased robustness to unknown poses [Savarese and Fei-Fei, 2008] and the use of dense feature representations [Su et al., 2009].

When combining detectors for multiple categories or viewpoints in a common system, an important consideration is how to perform the training and classifier evaluation steps efficiently. Torralba et al. [2004] propose a joint boosting approach for learning classifiers that share features between multiple classes, allowing for more efficient evaluation. Opelt et al. [2006b] also try to share spatial relations between classifiers for incremental learning of multiple object categories. Russovsky and Ng [2010] instead apply a segmentation approach to define regions-of-interest for efficient object detection and share segmentation parameters between different object classes in a multi-class setting. It is to be expected that the importance of feature and classifier sharing will further increase as recognition is scaled to larger numbers of categories.

12.4   ROLE OF VIDEO

As object detection progresses, it becomes amenable to more and more applications targeted at interpretation of continuous video data. In such applications, temporal continuity is an important cue that can be used to further stabilize the detection results (*e.g.*, [Everingham et al., 2006]), or to use them as the basis for tracking-by-detection approaches (*e.g.*, [Andriluka et al., 2008, 2010, Breitenstein et al., 2009, Choi and Savarese, 2010, Ess et al., 2009, Leibe et al., 2007, Leibe, Schindler and Van Gool, 2008, Okuma et al., 2004, Wu and Nevatia, 2007]). While such approaches naturally perform better with increasing detection quality, it is comforting to know that decent person and face tracking performance can already be achieved with the approaches presented in this tutorial.

12.5   INTEGRATED SEGMENTATION AND RECOGNITION

The idea to connect object recognition and class-specific segmentation has recently developed into an area of active research. In the following, we shortly review some main developments in this area, but an extensive treatment will need far more space.

When the goal is to recognize specific objects, segmentation can be seen as a dense refinement of the recognition process. Yu and Shi [2003] and Ferrari et al. [2004] both present simultaneous segmentation and recognition approaches for specific objects that follow this strategy. Yu & Shi formulate the segmentation problem in a graph theoretic framework that combines patch and pixel groupings, where the final solution is found using the Normalized Cuts criterion [Shi and Malik, 2000]. Ferrari *et al.* start from a small set of initial matches and then employ an iterative image exploration process that grows the matching region by searching for additional correspondences and segmenting the object in the process. Both methods achieve good segmentation results in cluttered real-world settings. However, both systems need to know the exact objects beforehand in order to extract their most discriminant features or search for additional correspondences.

For object category recognition, the task is again more difficult since exact correspondences between category members are generally not available. In many cases, reasonable bottom-up segmentations can already be obtained starting from the detection bounding boxes. Given an object bounding box, a set of pixels can usually be identified that clearly belong to the object (*e.g.*, from an elliptical prior region). Those pixels can be used to estimate the object's color distribution; other pixels that are clearly outside the object yield an estimate for the background distribution. Those two distributions can serve as input for a bottom-up segmentation procedure that tries to refine the object's contour, *e.g.*, using the popular GrabCut approach [Rother et al., 2004] based on a Conditional Random Field (CRF) formulation. Such a refinement step has been proposed by Ferrari et al. [2008] as an initialization for body pose estimation. As such an approach does not rely on detailed class-specific information for the segmentation process, it is widely applicable to different object categories and articulations. However, this also means that the segmentation process cannot take advantage of any specific knowledge of the target category. As a result, it runs the risk of obtaining incomplete segmentations, *e.g.*, cutting off a person's head or legs when the clothing colors are sufficiently different.

An alternative approach is to estimate a class-specific top-down segmentation based on the detection result. Borenstein and Ullman [2002] proposed an early approach to create class-specific segmentations by fitting local image fragments with associated figure-ground labels to the test image and combining them in jigsaw-puzzle fashion. As their approach, however, only considered local consistency between overlapping patches, the segmentation quality was quite limited.

As shown in Section 11.4.3, feature-based detectors such as the ISM or Hough Forests can be extended to yield a top-down segmentation as a result of the recognition process. This is achieved by back-projecting the votes contributing to a local maximum in the Hough space to the image in order to propagate top-down information to the patches they were originating from. This process can be used to infer local figure-ground labels Leibe, Leonardis and Schiele [2008], Leibe and Schiele [2003], object part annotations, or even depth maps or surface orientations from a single image Thomas et al. [2009*a*].

Naturally, the advantages of top-down and bottom-up segmentation processes should be combined. Several approaches have been proposed towards this goal. Currently, CRF formulations seem to be the most promising direction, as they allow to easily integrate class-specific per-pixel information with region-based information and neighborhood constraints [Kumar et al., 2005, Ladický et al., 2009, 2010, Larlus et al., 2008, Tu et al., 2003].

12.6 SUPERVISION CONSIDERATIONS IN OBJECT CATEGORY LEARNING

Given the success of statistical learning methods for building object models, an important question is how to assure adequate human supervision. Carefully prepared image datasets are valuable but expensive to obtain. Thus, researchers are exploring a variety of ways to loosen labeling requirements, minimize the amount of annotated data required, and even discover visual patterns without any direct supervision. We briefly note some emerging themes in these directions.

12.6.1 USING WEAKLY LABELED IMAGE DATA

Weakly supervised visual learning algorithms attempt to build category models from loosely or ambiguously labeled data. Most assume that each image contains a single prominent object of interest, and an image-level label records what object that is. Whereas "strong" annotations would mark the bounding box or segmentation for the object of interest, these labels are "weak" because they lack localization information. The general approach is to discover the common and/or discriminative portions of the weakly labeled exemplars for a given category. Robust models are then built to categorize *and* localize objects in novel images.

Perhaps most well-known are weak supervision paradigms developed for part-based models [Burl et al., 1998, Fergus et al., 2003, Weber et al., 2000*b*], which take an EM approach to simultaneously estimate parts and the model parameters (see Section 10.3.1).

A second strategy exploiting weakly labeled data is to simultaneously estimate the figure-ground segmentation of images from the same object category [Arora et al., 2007, Borenstein and Ullman, 2002, Cour and Shi, 2007, Kumar et al., 2005, Lee and Grauman, 2010a, Todorovic and Ahuja, 2006, Winn and Jojic, 2005, Yu et al., 2002]. In contrast to the above part-based strategy that extracts sparse patches based on their appearance and mutual positions, these approaches typically use regions produced from segmentation algorithms and contours produced from edge detection in order to pull out a *dense* connected area in each image. The segmentation-based methods are applicable for learning an object model, and also for the end goal of segmenting the input exemplar images.

A third general approach is to formulate the classification problem in the *multiple instance learning* (MIL) setting [Dietterich et al., 1997]. Whereas the above strategies attempt to discover the commonalities among weakly labeled exemplars, MIL methods instead treat the supervision as a form of labeling known to be noisy [Dollár et al., 2008, Maron and Ratan, 1998, Vijayanarasimhan and Grauman, 2008a, Viola et al., 2005, Yang and Lozano-Perez, 2000, Zhang and Goldman, 2002]. In contrast to the traditional supervised learning setting in which each instance has a binary label, in MIL, the learner is instead provided with *sets* (or bags) of points, and is only told that at least one member of any *positive bag* is truly positive, while every member of any *negative bag* is guaranteed to be negative. The goal of MIL is to induce the function that will accurately label individual instances such as the ones within the training bags in spite of the label ambiguity: the ratio of negative to positive instances within every positive bag can be arbitrarily high. Multi-label variants of MIL that are suitable for this supervision setting have been considered [Vijayanarasimhan and Grauman, 2009, Zha et al., 2008, Zhang and Zhou, 2007, Zhou and Zhang, 2006].

12.6.2 MAXIMIZING THE USE OF MANUAL ANNOTATIONS

Aside from accommodating noisy or weak labels, another line of work aims to most efficiently leverage minimal available labeled data. To this end, both *active learning* and *transfer learning* have begun to be explored for object recognition.

Active learning methods survey unlabeled data to select informative examples for which labels should be obtained from a human annotator. The classifiers are initialized with some labeled data, and then the system repeatedly forms label queries to feed to the annotators; given their reply, the classifiers are updated, and the cycle repeats. Active learners pinpoint the weaknesses in the model and thereby learn more efficiently than a traditional "passive" (random) labeling approach. Initial methods for active object category learning treat images as data points and devise uncertainty-based criteria to select images for labeling [Collins et al., 2008, Holub et al., 2008, Kapoor et al., 2007]. While selection functions are often easier to formulate in the binary class setting, given the importance of multi-class category learning for object recognition, vision researchers have also developed novel selection methods well-suited to multi-class data [Jain and Kapoor, 2009, Joshi et al., 2009]. The richness of image annotations also necessitates targeting wider types of requests to annotators, which calls for active selection criteria that weigh the examples and the possible questions to be asked in a unified manner. For example, the annotator may be asked to provide a bounding box around all objects of interest, or to label a specific region, or to provide manual segmentations on some image [Vijayanarasimhan and Grauman, 2008b, 2009]. Further extensions in this direction show how to generate questions about relationships and context [Siddiquie and Gupta, 2010].

A recognition system can also make do with less total training data by transferring knowledge of some categories to more efficiently learn others [Bart and Ullman, 2005, Fei-Fei et al., 2003, Lampert et al., 2009, Quattoni et al., 2008, Stark et al., 2009]. A common theme in object recognition transfer learning work is to identify the shared visual aspects of different object categories. One form of transfer learns object models from just a few images by incorporating "generic" knowledge obtained from previously learned models of unrelated categories [Fei-Fei et al., 2003, Miller et al., 2000, Stark et al., 2009]. Rather than use prior knowledge over the model parameters, another approach is to explicitly identify local visual features that can be shared across multiple categories Bart and Ullman [2005], or to exploit known class relationships to identify likely useful features or prototypes [Levi et al., 2004, Quattoni et al., 2008].

12.6.3 UNSUPERVISED OBJECT DISCOVERY

Some work considers how to mine unlabeled images to *discover* object categories. These methods are applicable to identify the themes and summarize the collection visually, or to connect the unsupervised discovery component with a labeling interface so that a human can more efficiently annotate the data.

Several authors have studied probabilistic clustering methods originally developed for text—such as probabilistic Latent Semantic Analysis (pLSA), Latent Dirichlet Analysis, and Hierarchical Dirichlet Processes—to discover the hidden mixture of visual themes ("topics") in a collection of image data [Fergus, Fei-Fei, Perona and Zisserman, 2005, Li et al., 2007, Liu and Chen, 2007, Russell et al., 2006, Sivic et al., 2005]. The main idea is to use feature co-occurrence patterns in the images ("documents") to recover the relatively few underlying distributions that best account for the data. An image containing instances of several categories is modeled as a mixture of topics. The models are typically applied to bag of visual words descriptors. Having discovered the topics, one can describe an image compactly based on the mixture of topics it contains, or gather those images with the most clear instances of a given topic.

Other approaches treat the unsupervised visual category learning task as an image clustering problem. The main challenges for such methods are to compute appropriate pairwise affinities between all pairs of unlabeled images such that semantically meaningful partitions can be obtained, and to isolate relevant features when dealing with multi-object or cluttered images. Thus, several techniques show how to iterate between computing the foreground region of interest (or set of local features) per image, and the clusters resulting from those foreground selections [Kim et al., 2008, Kim and Torralba, 2009, Lee and Grauman, 2008, 2009a,b]. The goal is to simultaneously partition the images into the primary repeated objects, while also discovering the portions of each image that can be most consistently clustered with the rest. Drawing on techniques from data mining, other work shows how to mine for common semi-local feature configurations in large image collections [Chum et al., 2009, Parikh et al., 2009, Philbin and Zisserman, 2008, Quack et al., 2007].

12.7   LANGUAGE, TEXT, AND IMAGES

Finally, another exciting direction in visual recognition is to draw on language understanding and text processing in conjunction with visual models. Researchers are exploring how text that naturally accompanies images—such as captioned photos, images embedded in Web pages, movies with scripts, or tagged images on photo-sharing websites—can be used to enable novel forms of learning and data collection.

One way to exploit tagged or captioned photos is to recover the correspondence (or "translation") between region descriptors and the keywords that appear alongside them [Duygulu et al., 2002], or to model the joint distribution of words and regions [Barnard et al., 2003, Lavrenko et al., 2003, Li et al., 2009, Monay and Gatica-Perez, 2003]. Recent work further shows how exploiting predicates (e.g., isAbove(A,B)) provides even stronger cues tying the two data views together [Gupta and Davis, 2008]. For the special case of named people or faces, more direct associations can be successfully learned [Berg et al., 2004, Cour et al., 2010, Everingham et al., 2006, Jie et al., 2009]. For example, in [Berg et al., 2004], news captions occurring alongside photographs are used to cluster faces occurring in the pictures, and natural language tools like named entity extraction can be used to parse the text meaningfully.

Another general strategy is to learn a new image representation (or similarly, distance function) that is bolstered by having observed many examples together with relevant text. To this end, variants of metric learning [Makadia et al., 2008, Qi et al., 2009], transfer learning [Quattoni et al., 2007], matrix factorization [Loeff and Farhadi, 2008], random field models [Bekkerman and Jeon, 2007], and projection-based methods [Blaschko and Lampert, 2008, Hardoon and Shawe-Taylor, 2003, Hwang and Grauman, 2010, Yakhnenko and Honavar, 2009] have all been explored.

Most recently, researchers are investigating how visual properties that are named in natural language (e.g., "spotted", "white", "flat") may be beneficial as mid-level cues for object recognition. These so-called *visual attributes* offer an intermediate representation between low-level image features and high-level categories [Berg et al., 2010, Branson et al., 2010, Farhadi et al., 2009, Hwang et al., 2011, Kumar et al., 2009, Lampert et al., 2009, Parikh and Grauman, 2011, Rohrbach et al., 2010, Wang et al., 2009, Wang and Mori, 2010]. Whereas traditional object detectors are built via supervised learning on image features, an attribute-based detector might first predict the presence of an array of visual properties, and then use the outputs of those models as features to an object classification layer. Attributes are attractive because they allow a recognition system to do much more than simply predict category labels. For example, attributes shared among different object categories can facilitate zero-shot learning of novel classes [Lampert et al., 2009], and they also allow a system to generate meaningful descriptions of unfamiliar objects [Farhadi et al., 2009].

CHAPTER 13

Conclusions

In this lecture, we have organized current techniques for object recognition, focusing primarily on the ideas that appear to have traction for instance matching and learned generic category models. Using the basic frameworks outlined above, it is now possible to perform reliable recognition for many types of specific objects and for a number of generic object categories. As indicated by the ongoing threads discussed in the previous chapter, challenges in visual recognition are quite broad, and the field continues to move forward at an exciting pace.

Bibliography

Agarwal, A. and Triggs, B. [2006], Hyperfeatures multilevel local coding for visual recognition, *in* 'Proceedings of the European Conference on Computer Vision'. DOI: 10.1007/11744023_3 68, 69, 97, 98

Allan, M. and Williams, C. [2009], 'Object localization using the generative template of features', *Computer Vision and Image Understanding* **113**, 824–838. DOI: 10.1016/j.cviu.2009.02.002 85

Amit, Y., Geman, D. and Fan, X. [2004], 'A coarse-to-fine strategy for multi-class shape detection', *IEEE Transactions on Pattern Analysis and Machine Intelligence*. DOI: 10.1109/TPAMI.2004.111 80

Andriluka, M., Roth, S. and Schiele, B. [2008], People tracking-by-detection and people detection-by-tracking, *in* 'Proceedings of the IEEE Conference on Computer Vision and Pattern Recognition'. DOI: 10.1109/CVPR.2008.4587583 76, 115, 124

Andriluka, M., Roth, S. and Schiele, B. [2010], Monocular 3D pose estimation and tracking by detection, *in* 'Proceedings of the IEEE Conference on Computer Vision and Pattern Recognition'. DOI: 10.1109/CVPR.2010.5540156 124

Arora, H., Loeff, N., Forsyth, D. and Ahuja, N. [2007], Unsupervised segmentation of objects using efficient learning, *in* 'Proceedings of the IEEE Conference on Computer Vision and Pattern Recognition'. DOI: 10.1109/CVPR.2007.383011 126

Arya, S., Mount, D., Netanyahu, N., Silverman, R. and Wu, A. [1998], 'An optimal algorithm for approximate nearest neighbor searching in fixed dimensions', *Journal of the ACM* **45**, 891–923. DOI: 10.1145/293347.293348 29

Athitsos, V., Alon, J., Sclaroff, S. and Kollios, G. [2004], BoostMap: A method for efficient approximate similarity rankings, *in* 'Proceedings of the IEEE Conference on Computer Vision and Pattern Recognition'. DOI: 10.1109/CVPR.2004.1315173 33

Bach, F. R., Lanckriet, G. R. G. and Jordan, M. I. [2004], Fast kernel learning using sequential minimal optimization, Technical Report UCB/CSD-04-1307, EECS Department, University of California, Berkeley. 99

Baeza-Yates, R. and Ribeiro-Neto, B. [1999], *Modern Information Retrieval*, ACM Press. 36

Ballard, D. [1981], 'Generalizing the Hough Transform to detect arbitrary shapes', *Pattern Recognition* **13**(2), 111–122. DOI: 10.1016/0031-3203(81)90009-1 51

Bao, Y., Sun, M. and Savarese, S. [2010], Toward coherent object detection and scene layout understanding, *in* 'Proceedings of the IEEE Conference on Computer Vision and Pattern Recognition'. DOI: 10.1109/CVPR.2010.5540229 122

Barnard, K., Duygulu, P., de Freitas, N., Forsyth, D., Blei, D. and Jordan, M. [2003], 'Matching words and pictures', *Journal of Machine Learning Research* **3**, 1107–1135. DOI: 10.1162/153244303322533214 128

Barsalou, L. [1983], 'Ad-hoc categories', *Memory and Cognition* **11**, 211–227. 2

Bart, E. and Ullman, S. [2005], Cross-Generalization: Learning novel classes from a single example by feature replacement, *in* 'Proceedings of the IEEE Conference on Computer Vision and Pattern Recognition'. DOI: 10.1109/CVPR.2005.117 127

Bay, H., Ess, A., Tuytelaars, T. and Van Gool, L. [2008], 'SURF: Speeded-Up Robust Features', *Computer Vision and Image Understanding* **110**(3), 346–359. DOI: 10.1016/j.cviu.2007.09.014 xv, 24

Bay, H., Tuytelaars, T. and Van Gool, L. [2006], SURF: Speeded-Up Robust Features, *in* 'Proceedings of the European Conference on Computer Vision'. DOI: 10.1016/j.cviu.2007.09.014 xvi, 9, 24, 65

Beaudet, P. [1978], Rotationally invariant image operators, *in* 'Proc. 4th International Joint Conference on Pattern Recognition', pp. 579–583. 13

Beis, J. and Lowe, D. [1997], Shape indexing using approximate nearest-neighbour search in high dimensional spaces, *in* 'Proceedings of the IEEE Conference on Computer Vision and Pattern Recognition'. DOI: 10.1109/CVPR.1997.609451 28, 29

Bekkerman, R. and Jeon, J. [2007], Multi-modal clustering for multimedia collections, *in* 'Proceedings of the IEEE Conference on Computer Vision and Pattern Recognition'. DOI: 10.1109/CVPR.2007.383223 129

Belhumeur, P. and Kriegman, D. [1996], Eigenfaces vs. Fisherfaces: Recognition using class specific linear projection, *in* 'Proceedings of the European Conference on Computer Vision'. DOI: 10.1109/34.598228 7, 8

Belongie, S., Malik, J. and Puzicha, J. [2002], 'Shape matching and object recognition using shape contexts', *IEEE Transactions on Pattern Analysis and Machine Intelligence* **24**(24), 509–522. DOI: 10.1109/34.993558 69, 90, 95

Berg, A., Berg, T. and Malik, J. [2005], Shape matching and object recognition low distortion correspondences, *in* 'Proceedings of the IEEE Conference on Computer Vision and Pattern Recognition'. DOI: 10.1109/CVPR.2005.320 xvi, 95, 96

Berg, T., Berg, A., Edwards, J. and Forsyth, D. [2004], Who's in the picture?, *in* 'Advances in Neural Information Processing Systems'. 128

Berg, T., Berg, A. and Shih, J. [2010], Automatic attribute discovery and characterization from noisy web data, *in* 'Proceedings of the European Conference on Computer Vision'. 129

Besl, P. and Jain, R. [1895], Three dimensional object recognition, ACM Computing Surveys, **17**(1), March. DOI: 10.1145/4078.4081 xv, 5

Biederman, I. [1981], On the semantics of a glance at a scene, *in* M. Kubovy and J. Pomerantz, eds, 'Perceptual Organization', Lawrence Erlbaum, chapter 8. 122

Biederman, I. [1987], 'Recognition by components: A theory of human image understanding', *Psychology Review* **94**(2), 115–147. DOI: 10.1037/0033-295X.94.2.115 89

Blaschko, M. B. and Lampert, C. H. [2008], Correlational spectral clustering, *in* 'Proceedings of the IEEE Conference on Computer Vision and Pattern Recognition'. DOI: 10.1109/CVPR.2008.4587353 129

Bombini, L., Cerri, P., Grisleri, P., Scaffardi, S. and Zani, P. [2006], An evaluation of monocular image stabilization algorithms for automotive applications, *in* 'IEEE Int'l Conf. Intelligent Transportation Systems'. 122

Borenstein, E. and Ullman, S. [2002], Class-specific, top-down segmentation, *in* 'Proceedings of the European Conference on Computer Vision'. DOI: 10.1007/3-540-47967-8_8 125, 126

Bosch, A., Zisserman, A. and Munoz, X. [2007*a*], Image classification using random forests and ferns, *in* 'Proceedings of the IEEE International Conference on Computer Vision'. DOI: 10.1109/ICCV.2007.4409066 37

Bosch, A., Zisserman, A. and Munoz, X. [2007*b*], Representing shape with a spatial pyramid kernel, *in* 'ACM International Conference on Image and Video Retrieval'. DOI: 10.1145/1282280.1282340 xvi, 64, 65

Bouchard, G. and Triggs, B. [2005], Hierarchical part-based visual object categorization, *in* 'Proceedings of the IEEE Conference on Computer Vision and Pattern Recognition'. DOI: 10.1109/CVPR.2005.174 72

Bourdev, L. and Malik, J. [2009], Poselets: body part detectors trained using 3D human pose annotations, *in* 'Proceedings of the IEEE International Conference on Computer Vision'. DOI: 10.1109/ICCV.2009.5459303 77

Boureau, Y.-L., Bach, F., LeCun, Y. and Ponce, J. [2010], Learning mid-level features for recognition, *in* 'Proceedings of the IEEE Conference on Computer Vision and Pattern Recognition'. DOI: 10.1109/CVPR.2010.5539963 109

Branson, S., Wah, C., Babenko, B., Schroff, F., Welinder, P., Belongie, S., and Perona, P. [2010], Visual recognition with humans in the loop, *in* 'Proceedings of the European Conference on Computer Vision. 129

Breitenstein, M., Reichlin, F., Leibe, B., Koller-Meier, E. and Van Gool, L. [2009], Robust tracking-by-detection using a detector confidence particle filter, *in* 'Proceedings of the IEEE International Conference on Computer Vision'. DOI: 10.1109/ICCV.2009.5459278 124

Broder, A. [1998], On the resemblance and containment of documents, *in* 'Compression and Complexity of Sequences'. DOI: 10.1109/SEQUEN.1997.666900 32

Brown, M. and Lowe, D. [2002], Invariant features from interest point groups, *in* 'British Machine Vision Conference', Cardiff, Wales, pp. 656–665. 18

Brown, M. and Lowe, D. [2003], Recognising panoramas, *in* 'Proceedings of the IEEE International Conference on Computer Vision'. DOI: 10.1109/ICCV.2003.1238630 xv, 55, 56

Brown, M. and Lowe, D. [2007], 'Automatic panoramic image stitching using invariant features', *International Journal of Computer Vision* **74**(1), 59–73. DOI: 10.1007/s11263-006-0002-3 xv, 55, 56

Brown, R. [1958], 'How shall a thing be called?', *Psychological Review* **65**, 14–21. DOI: 10.1037/h0041727 1

Burl, M., Weber, M. and Perona, P. [1998], A probabilistic approach to object recognition using local photometry and gl obal geometry, *in* 'Proceedings of the European Conference on Computer Vision'. 126

Capel, D. [2005], An effective bail-out test for RANSAC consensus scoring, *in* 'British Machine Vision Conference', pp. 629–638. 51

Carlsson, S. [1998], Order structure, correspondence and shape based categories, *in* 'Intl Wkshp on Shape Contour and Grouping', Sicily. 90

Carneiro, G. and Lowe, D. [2006], Sparse flexible models of local features, *in* 'Proceedings of the European Conference on Computer Vision'. xvi, 71, 72

Charikar, M. [2002], Similarity estimation techniques from rounding algorithms, *in* 'ACM Symp. on Theory of Computing'. DOI: 10.1145/509907.509965 31, 32

Choi, J., Jeon, W. and Lee, S.-C. [2008], Spatio-temporal pyramid matching for sports videos, *in* 'Proceeding of the ACM International Conference on Multimedia Information Retrieval', Vancouver, Canada. DOI: 10.1145/1460096.1460144 94

Choi, W. and Savarese, S. [2010], Multiple target tracking in world coordinate with single, minimally calibrated camera, *in* 'Proceedings of the European Conference on Computer Vision'. DOI: 10.1007/978-3-642-15561-1_40 124

Chopra, S., Hadsell, R. and LeCun, Y. [2005], Learning a similarity metric discriminatively, with Application to Face Verification, *in* 'Proceedings of the IEEE Conference on Computer Vision and Pattern Recognition', San Diego, CA. DOI: 10.1109/CVPR.2005.202 99

Chui, H. and Rangarajan, A. [2000], A new algorithm for non-rigid point matching, *in* 'Proceedings of the IEEE Conference on Computer Vision and Pattern Recognition', Hilton Head Island, SC. DOI: 10.1109/CVPR.2000.854733 90, 95

Chum, O. and Matas, J. [2005], Matching with PROSAC - Progressive Sample Consensus, *in* 'Proceedings of the IEEE Conference on Computer Vision and Pattern Recognition'. DOI: 10.1109/CVPR.2005.221 51

Chum, O. and Matas, J. [2008], 'Optimal randomized RANSAC', *IEEE Transactions on Pattern Analysis and Machine Intelligence* **30**(8), 1472–1482. DOI: 10.1109/TPAMI.2007.70787 51

Chum, O., Matas, J. and Obdržálek, S. [2004], Enhancing RANSAC by generalized model optimization, *in* 'Asian Conference on Computer Vision', pp. 812–817. 51

Chum, O., Perdoch, M. and Matas, J. [2009], Geometric min-Hashing: Finding a (thick) needle in a haystack, *in* 'Proceedings of the IEEE Conference on Computer Vision and Pattern Recognition'. 128

Chum, O., Philbin, J. and Zisserman, A. [2008], Near duplicate image detection: min-hash and tf-idf weighting, *in* 'British Machine Vision Conference'. 32

Chum, O., Werner, T. and Matas, J. [2005], Two-view geometry estimation unaffected by a dominant plane, *in* 'Proceedings of the IEEE Conference on Computer Vision and Pattern Recognition', pp. 772–779. DOI: 10.1109/CVPR.2005.354 51

Chum, O. and Zisserman, A. [2007], An exemplar model for learning object classes, *in* 'Proceedings of the IEEE Conference on Computer Vision and Pattern Recognition'. DOI: 10.1109/CVPR.2007.383050 77, 94

Ciaccia, P., Patella, M. and Zezula, P. [1997], M-tree: An efficient access method for similarity search in metric spaces, *in* 'Proc Int'l Conf on Very Large Data Bases'. 29

Collins, B., Deng, J., Li, K. and Fei-Fei, L. [2008], Towards scalable dataset construction: An active learning approach., *in* 'Proceedings of the European Conference on Computer Vision'. DOI: 10.1007/978-3-540-88682-2_8 127

Comaniciu, D. and Meer, P. [2002], 'Mean shift: A robust approach toward feature space analysis', *IEEE Transactions on Pattern Analysis and Machine Intelligence* **24**(5), 603–619. DOI: 10.1109/34.1000236 84

Cootes, T., Edwards, G. and Taylor, C. [2001], 'Active appearance models', *IEEE Transactions on Pattern Analysis and Machine Intelligence* **23**(6). xvi, 89, 91

Cornelis, N. and Van Gool, L. [2008], Fast scale invariant feature detection and matching on programmable graphics hardware, *in* 'IEEE CVPR Workshop on Computer Vision on the GPU', Anchorage, USA. DOI: 10.1109/CVPRW.2008.4563087 24, 25

Cour, T., Sapp, B., Nagle, A. and Taskar, B. [2010], Talking pictures: temporal grouping and dialog-supervised person recognition, *in* 'Proceedings of the IEEE Conference on Computer Vision and Pattern Recognition'. DOI: 10.1109/CVPR.2010.5540106 128

Cour, T. and Shi, J. [2007], Recognizing objects by piecing together the segmentation puzzle, *in* 'Proceedings of the IEEE Conference on Computer Vision and Pattern Recognition'. DOI: 10.1109/CVPR.2007.383051 126

Crandall, D., Felzenszwalb, P. and Huttenlocher, D. [2005], Spatial priors for part-based recognition using statistical models, *in* 'Proceedings of the IEEE Conference on Computer Vision and Pattern Recognition'. DOI: 10.1109/CVPR.2005.329 72

Cristianini, N., Shawe-Taylor, J. and Elisseeff, A. [2001], On kernel-target alignment, *in* 'Advances in Neural Information Processing Systems'. DOI: 10.1007/3-540-33486-6_8 99

Csurka, G., Bray, C., Dance, C. and Fan, L. [2004], Visual categorization with bags of keypoints, *in* 'Workshop on Statistical Learning in Computer Vision, in conjunction with ECCV'. 64, 68, 108

Cula, O. and Dana, K. [2001], Compact representation of bidirectional texture functions, *in* 'Proceedings of the IEEE Conference on Computer Vision and Pattern Recognition'. DOI: 10.1109/CVPR.2001.990645 68

Dalal, N. and Triggs, B. [2005], Histograms of oriented gradients for human detection, *in* 'Proceedings of the IEEE Conference on Computer Vision and Pattern Recognition'. DOI: 10.1109/CVPR.2005.177 64, 65, 76, 77, 81, 107, 115

Datar, M., Immorlica, N., Indyk, P. and Mirrokni, V. [2004], Locality-sensitive hashing scheme based on p-stable distributions, *in* 'Symposium on Computational Geometry (SOCG)'. DOI: 10.1145/997817.997857 31, 32

De Graef, P., Christiaens, D. and d'Ydewalle, G. [1990], 'Perceptual effects of scene context on object identification', *Psychological Research* **52**, 317–329. DOI: 10.1007/BF00868064 122

Deng, J., Dong, W., Socher, R., Li, L.-J., Li, K. and Fei-Fei, L. [2009], ImageNet: A large-scale hierarchical image database, *in* 'CVPR09'. DOI: 10.1109/CVPR.2009.5206848 121

Desai, C., Ramanan, D. and Fowlkes, C. [2009], Discriminative models for multi-class object layout, *in* 'Proceedings of the IEEE International Conference on Computer Vision'. DOI: 10.1109/ICCV.2009.5459256 123

Dietterich, T., Lathrop, R. and Lozano-Perez, T. [1997], 'Solving the multiple instance problem with axis-parallel rectangles', *Artificial Intelligence* **89**(1-2), 31–71. DOI: 10.1016/S0004-3702(96)00034-3 126

Dollár, P., Babenko, B., Belongie, S., Perona, P. and Tu, Z. [2008], Multiple component learning for object detection, *in* 'ECCV'. DOI: 10.1007/978-3-540-88688-4_16 126

Dorko, G. and Schmid, C. [2003], Selection of scale-invariant parts for object class recognition, *in* 'Proceedings of the IEEE International Conference on Computer Vision'. DOI: 10.1109/ICCV.2003.1238407 70

Duygulu, P., Barnard, K., de Freitas, N. and Forsyth, D. [2002], Object recognition as machine translation: learning a lexicon for a fixed image vocabulary, *in* 'Proceedings of the European Conference on Computer Vision'. DOI: 10.1007/3-540-47979-1_7 128

Ess, A., Leibe, B., Schindler, K. and Van Gool, L. [2009], 'Robust multi-person tracking from a mobile platform', *IEEE Transactions on Pattern Analysis and Machine Intelligence* **31**(10), 1831–1846. DOI: 10.1109/TPAMI.2009.109 124

Ess, A., Leibe, B. and Van Gool, L. [2007], Depth and appearance for mobile scene analysis, *in* 'Proceedings of the IEEE International Conference on Computer Vision'. DOI: 10.1109/ICCV.2007.4409092 122

Everingham, M., Sivic, J. and Zisserman, A. [2006], "Hello! My name is... Buffy" – Automatic naming of characters in TV video, *in* 'British Machine Vision Conference'. 124, 128

Everingham, M., Van Gool, L., Williams, C., Winn, J. and Zisserman, A. [2008], 'The PASCAL visual object classes challenge 2008 (VOC2008) results', http://www.pascal-network.org/challenges/VOC/voc2008/workshop/index.html. 68, 77, 87, 119, 120

Everingham, M., Van Gool, L., Williams, C., Winn, J. and Zisserman, A. [2010], 'The Pascal visual object classes (VOC) challenge', *International Journal of Computer Vision* **88**(2), 303–338. DOI: 10.1007/s11263-009-0275-4 101, 115, 116, 119

Fan, T.-J., Medioni, G., and Nevatia, R. [1989], Recognizing 3-D objects using surface descriptions, *in IEEE Transactions on Pattern Analysis and Machine Intelligence*, **11**(11), 1989. DOI: 10.1109/34.42853 xv, 5

Farhadi, A., Endres, I., Hoiem, D. and Forsyth, D. [2009], Describing objects by their attributes, *in* 'Proceedings of the IEEE Conference on Computer Vision and Pattern Recognition'. DOI: 10.1109/CVPRW.2009.5206772 129

Fei-Fei, L., Fergus, R. and Perona, P. [2003], A Bayesian approach to unsupervised one-shot learning of object categories, *in* 'Proceedings of the IEEE International Conference on Computer Vision'. DOI: 10.1109/ICCV.2003.1238476 72, 77, 127

Fei-Fei, L., Fergus, R. and Perona, P. [2004], Learning generative visual models from few training examples: an incremental Bayesian approach tested on 101 object categories, *in* 'Workshop on Generative Model Based Vision'. DOI: 10.1016/j.cviu.2005.09.012 68

Felleman, D. J. and van Essen, D. C. [1991], 'Distributed hierarchical processing in the primate cerebral cortex', *Cerebral Cortex* **1**, 1–47. DOI: 10.1093/cercor/1.1.1-a 3

Felzenszwalb, P., Girshick, R. and McAllester, D. [2010], Cascade object detection with deformable part models, *in* 'Proceedings of the IEEE Conference on Computer Vision and Pattern Recognition'. 118

Felzenszwalb, P., Girshick, R., McAllester, D. and Ramanan, D. [2010], 'Object detection with discriminatively trained part based models', *IEEE Transactions on Pattern Analysis and Machine Intelligence* **32**(9). DOI: 10.1109/TPAMI.2009.167 xvii, 72, 101, 102, 115, 116, 117, 123

Felzenszwalb, P. and Huttenlocher, D. [2005], 'Pictorial structures for object recognition', *International Journal of Computer Vision* **61**(1). DOI: 10.1023/B:VISI.0000042934.15159.49 72, 76, 86, 91, 115, 116

Felzenszwalb, P., McAllester, D. and Ramanan, D. [2008], A discriminatively trained, multiscale, deformable part model, *in* 'Proceedings of the IEEE Conference on Computer Vision and Pattern Recognition'. DOI: 10.1109/CVPR.2008.4587597 72, 76, 77, 101, 102, 115, 117

Fergus, R., Fei-Fei, L., Perona, P. and Zisserman, A. [2005], Learning object categories from Google's image search, *in* 'Proceedings of the IEEE International Conference on Computer Vision'. DOI: 10.1109/ICCV.2005.142 128

Fergus, R., Perona, P. and Zisserman, A. [2005], A sparse object category model for efficient learning and exhaustive recognition, *in* 'Proceedings of the IEEE Conference on Computer Vision and Pattern Recognition'. DOI: 10.1109/CVPR.2005.47 72, 77

Fergus, R., Zisserman, A. and Perona, P. [2003], Object class recognition by unsupervised scale-invariant learning, *in* 'Proceedings of the IEEE Conference on Computer Vision and Pattern Recognition'. DOI: 10.1109/CVPR.2003.1211479 xvi, 72, 73, 75, 100, 126

Ferrari, V., Marin, M. and Zisserman, A. [2008], Progressive search space reduction for human pose estimation, *in* 'Proceedings of the IEEE Conference on Computer Vision and Pattern Recognition'. DOI: 10.1109/CVPR.2008.4587468 76, 125

Ferrari, V., Tuytelaars, T. and Gool, L. V. [2006], Object detection by contour segment networks, *in* 'Proceedings of the European Conference on Computer Vision'. DOI: 10.1007/11744078_2 69

Ferrari, V., Tuytelaars, T. and van Gool, L. [2004], Simultaneous recognition and segmentation by image exploration, *in* 'Proceedings of the European Conference on Computer Vision'. DOI: 10.1109/ICIP.2010.5654176 124

Fischler, M. and Bolles, R. [1981], 'Random sampling consensus: A paradigm for model fitting with application to image analysis and automated cartography', *Communications of the ACM* **24**, 381–395. DOI: 10.1145/358669.358692 48

Fischler, M. and Elschlager, R. [1973], 'The representation and matching of pictorial structures', *IEEE Transactions on Computers* **22**(1), 67–92. DOI: 10.1109/T-C.1973.223602 71, 85

Fleuret, F. and Geman, D. [2001], Coarse-to-fine face detection, *in* 'International Journal of Computer Vision'. DOI: 10.1023/A:1011113216584 80, 81

Förstner, W. and Gülch, E. [1987], A fast operator for detection and precise location of distinct points, corners and centres of circular features, *in* 'ISPRS Intercommission Workshop', Interlaken. 13

Frahm, J.-M. and Pollefeys, M. [2006], RANSAC for (quasi-) degenerate data (QDEGSAC), *in* 'Proceedings of the IEEE Conference on Computer Vision and Pattern Recognition', pp. 453–460. DOI: 10.1109/CVPR.2006.235 51

Freeman, W. and Roth, M. [1995], Orientation histograms for hand gesture recognition, *in* 'International Workshop on Automatic Face and Gesture Recognition'. 107

Friedman, J., Bentley, J. and Finkel, A. [1977], 'An algorithm for finding best matches in logarithmic expected time', *ACM Transactions on Mathematical Software* **3**(3), 209–226. DOI: 10.1145/355744.355745 28

Freund, Y. and Schapire, R. [1995], A decision-theoretic generalization of online learning and an application to boosting, *in* 'Computational Learning Theory'. DOI: 10.1006/jcss.1997.1504 104

Fritz, M., Leibe, B., Caputo, B. and Schiele, B. [2005], Integrating representative and discriminant models for object category detection, *in* 'Proceedings of the IEEE International Conference on Computer Vision'. DOI: 10.1109/ICCV.2005.124 115

Frome, A., Singer, Y. and Malik, J. [2007], Image retrieval and classification using local distance functions, *in* 'Advances in Neural Information Processing Systems 19'. 98

Frome, A., Singer, Y., Sha, F. and Malik, J. [2007], Learning globally-consistent local distance functions for shape-based image retrieval and classification, *in* 'Proceedings of the IEEE International Conference on Computer Vision'. DOI: 10.1109/ICCV.2007.4408839 98

Gall, J. and Lempitsky, V. [2009], Class-specific hough forests for object detection, *in* 'Proceedings of the IEEE Conference on Computer Vision and Pattern Recognition'. DOI: 10.1109/CVPRW.2009.5206740 77, 115, 123

Galleguillos, C., Rabinovich, A. and Belongie, S. [2008], Object categorization using co-occurrence, location and appearance, *in* 'Proceedings of the IEEE Conference on Computer Vision and Pattern Recognition'. DOI: 10.1109/CVPR.2008.4587799 81, 123

Gammeter, S., Bossard, L., Quack, T. and Van Gool, L. [2009], I know what you did last summer: object-level auto-annotation of holiday snaps, *in* 'Proceedings of the IEEE International Conference on Computer Vision'. xvi, 58, 59

Gavrila, D. and Munder, S. [2007], 'Multi-cue pedestrian detection and tracking from a moving vehicle', *International Journal of Computer Vision* **73**(1), 41–59. DOI: 10.1007/s11263-006-9038-7 122

Gavrila, D. and Philomin, V. [1999], Real-time object detection for smart vehicles, *in* 'Proceedings of the IEEE International Conference on Computer Vision'. DOI: 10.1109/ICCV.1999.791202 69

Geronimo, D., Sappa, A., Ponsa, D. and Lopez, A. [2010], '2D-3D-based on-board pedestrian detection system', *Computer Vision and Image Understanding* **114**(5), 583–595. DOI: 10.1016/j.cviu.2009.07.008 122

Gionis, A., Indyk, P. and Motwani, R. [1999], Similarity search in high dimensions via hashing, *in* 'Proceedings of the 25th International Conference on Very Large Data Bases'. 31, 32

Gold, S. and Rangarajan, A. [1996], 'A graduated assignment algorithm for graph matching', *IEEE Transactions on Pattern Analysis and Machine Intelligence* **18**(4), 377–388. DOI: 10.1109/34.491619 90, 95

GPU [2008], 'GPUSURF features website', http://homes.esat.kuleuven.be/~ncorneli/gpusurf/. 25

Grauman, K. and Darrell, T. [2005], The pyramid match kernel: Discriminative classification with sets of image features, *in* 'Proceedings of the IEEE International Conference on Computer Vision'. DOI: 10.1109/ICCV.2005.239 37, 91, 92

Grauman, K. and Darrell, T. [2006*a*], Approximate correspondences in high dimensions, *in* 'Advances in Neural Information Processing Systems'. xvi, 37, 90, 93

Grauman, K. and Darrell, T. [2006*b*], Unsupervised learning of categories from sets of partially matching image features, *in* 'Proceedings of the IEEE Conference on Computer Vision and Pattern Recognition'. DOI: 10.1109/CVPR.2006.322

Grauman, K. and Darrell, T. [2007*a*], Pyramid match hashing: Sub-linear time indexing over partial correspondences, *in* 'Proceedings of the IEEE Conference on Computer Vision and Pattern Recognition'. DOI: 10.1109/CVPR.2007.383225 32

Grauman, K. and Darrell, T. [2007*b*], 'The pyramid match kernel: Efficient learning with sets of features', *Journal of Machine Learning Research (JMLR)* **8**, 725–760. 91

Grauman, K. and Leibe, B. [2008], *AAAI 2008 tutorial on visual recognition.* xi

Griffin, G., Holub, A. and Perona, P. [2007], Caltech-256 object category dataset, Technical Report 7694, California Institute of Technology. 87, 119, 120

Grimson, W.E.L. and Huttenlocher, D.P., [1991], On the verification of the hypothesized matches in model-based recognition, *IEEE Transactions on Pattern Analysis and Machine Intelligence* **13**(12), 1201–1213. DOI: 10.1109/34.106994 xv, 5

Gupta, A. and Davis, L. [2008], Beyond nouns: Exploiting prepositions and comparative adjectives for learning visual classifiers, *in* 'Proceedings of the European Conference on Computer Vision'. DOI: 10.1007/978-3-540-88682-2_3 128

Hardoon, D. and Shawe-Taylor, J. [2003], KCCA for different level precision in content-based image retrieval, *in* 'Third International Workshop on Content-Based Multimedia Indexing'. 129

Harris, C. and Stephens, M. [1988], A combined corner and edge detector, *in* 'Alvey Vision Conference', pp. 147–151. 13, 67

Hartley, R. and Zisserman, A. [2004], *Multiple View Geometry in Computer Vision*, 2nd edn, Cambridge Univ. Press. 46, 47, 48

He, X., Zemel, R. and Carreira-Perpinan, M. [2004], Multiscale conditional random fields for image labeling, *in* 'Proceedings of the IEEE Conference on Computer Vision and Pattern Recognition'. DOI: 10.1109/CVPR.2004.1315232 81, 122

Hedau, V., Hoiem, D. and Forsyth, D. [2009], Recovering the spatial layout of cluttered rooms, *in* 'Proceedings of the IEEE International Conference on Computer Vision'. 122

Heisele, B., Serre, T., Pontil, M. and Poggio, T. [2001], Component-based face detection, *in* 'Proceedings of the IEEE Conference on Computer Vision and Pattern Recognition', pp. 657–662. DOI: 10.1109/CVPR.2001.990537 83, 101

Heitz, G. and Koller, D. [2008], Learning spatial context: Using stuff to find things, *in* 'Proceedings of the European Conference on Computer Vision'. DOI: 10.1007/978-3-540-88682-2_4 81, 123

Hertz, T., Bar-Hillel, A. and Weinshall, D. [2004], Learning distance functions for image retrieval, *in* 'Proceedings of the IEEE Conference on Computer Vision and Pattern Recognition'. DOI: 10.1109/CVPR.2004.1315215 99

Hoiem, D., Efros, A. and Hebert, M. [2006], Putting objects into perspective, *in* 'Proceedings of the IEEE Conference on Computer Vision and Pattern Recognition'. DOI: 10.1109/CVPR.2006.232 82, 122

Hoiem, D., Rother, C. and Winn, J. [2007], 3D layout CRF for multi-view object class recognition and segmentation, *in* 'Proceedings of the IEEE Conference on Computer Vision and Pattern Recognition'. 124

Holub, A., Burl, M. and Perona, P. [2008], Entropy-based active learning for object recognition, *in* 'IEEE Workshop on Online Learning for Classification'. DOI: 10.1109/CVPRW.2008.4563068 127

Holub, A., Liu, Y. and Perona, P. [2007], On constructing facial similarity maps, *in* 'Proceedings of the IEEE Conference on Computer Vision and Pattern Recognition'. DOI: 10.1109/CVPR.2007.383281 99

Hough, P. [1962], 'Method and means for recognizing complex patterns', U.S. Patent 3069654. 51

Hu, Y., Li, M., Li, Z. and Ma, W.-Y. [2006], *Advances in multimedia modeling*, Vol. 4351, Springer, chapter Dual-Space Pyramid Matching for Medical Image Classification, pp. 96–105. 94

Huang, J., Kumar, S., Mitra, M., Zhu, W. and Zabih, R. [1999], 'Color-spatial indexing and applications', *International Journal of Computer Vision* **35**(3), 245–268. DOI: 10.1023/A:1008108327226 98

Hubel, D. and Wiesel, T. [1959], 'Receptive fields of single neurons in the cat's striate cortex', *Journal of Physiology* **148**, 574–591. 64

Hubel, D. and Wiesel, T. [1977], 'Functional architecture of macaque monkey visual cortex', *Proc. of the Royal Society B Biological Sciences* **198**, 1–59. DOI: 10.1098/rspb.1977.0085 64

Hwang, S. J. and Grauman, K. [2010], Accounting for the relative importance of objects in image retrieval, *in* 'British Machine Vision Conference'. DOI: 10.5244/C.24.58 129

Hwang, S. J., Sha, F., and Grauman, K. [2011], 'Sharing features between objects and their attributes', *in* 'Proceedings of the IEEE Conference on Computer Vision and Pattern Recognition'. 129

Indyk, P. and Motwani, R. [1998], Approximate nearest neighbors: Towards removing the curse of dimensionality, *in* '30th Symposium on Theory of Computing'. DOI: 10.1145/276698.276876 31, 32

Jain, P. and Kapoor, A. [2009], Active learning for large multi-class problems, *in* 'Proceedings of the IEEE Conference on Computer Vision and Pattern Recognition'. DOI: 10.1109/CVPRW.2009.5206651 127

Jain, P., Kulis, B. and Grauman, K. [2008*a*], Fast image search for learned metrics, *in* 'Proceedings of the IEEE Conference on Computer Vision and Pattern Recognition'. DOI: 10.1109/CVPR.2008.4587841 33, 99

Jain, P., Kulis, B., Dhillon, I., and Grauman, K. [2008*b*], Online metric learning and fast similarity search, *in* 'Advances in Neural Information Processing Systems'. 33

Jie, L., Caputo, B. and Ferrari, V. [2009], Who's doing what: Joint modeling of names and verbs for simultaneous face and pose annotation, *in* 'Advances in Neural Information Processing Systems'. 128

Johnson, A. and Hebert, M. [1999], 'Using spin images for efficient object recognition in cluttered 3D scenes', *IEEE Transactions on Pattern Analysis and Machine Intelligence* **21**(5), 433–449. DOI: 10.1109/34.765655 90

Jones, M. and Rehg, J. [1999], 'Statistical color models with application to skin detection', *International Journal of Computer Vision*. DOI: 10.1023/A:1013200319198 63

Joshi, A., Porikli, F. and Papanikolopoulos, N. [2009], Multi-class active learning for image classification, *in* 'Proceedings of the IEEE Conference on Computer Vision and Pattern Recognition'. DOI: 10.1109/CVPRW.2009.5206627 127

Kadir, T. and Brady, M. [2001], 'Scale, saliency, and image description', *International Journal of Computer Vision* **45**(2), 83–105. DOI: 10.1023/A:1012460413855 21

Kadir, T., Zisserman, A. and Brady, M. [2004], An affine invariant salient region detector, *in* 'Proceedings of the European Conference on Computer Vision'. DOI: 10.1007/978-3-540-24670-1_18 21

Kapoor, A., Grauman, K., Urtasun, R. and Darrell, T. [2007], Active learning with Gaussian processes for object categorization, *in* 'Proceedings of the IEEE International Conference on Computer Vision'. DOI: 10.1109/ICCV.2007.4408844 127

Kim, G., Faloutsos, C. and Hebert, M. [2008], Unsupervised modeling of object categories using link analysis techniques, *in* 'Proceedings of the IEEE Conference on Computer Vision and Pattern Recognition'. DOI: 10.1109/CVPR.2008.4587502 128

Kim, G. and Torralba, A. [2009], Unsupervised detection of regions of interest using link analysis, *in* 'Advances in Neural Information Processing Systems'. 128

Kuhn, H. [1955], 'The Hungarian method for the assignment problem', *Naval Research Logistic Quarterly* **2**, 83–97. DOI: 10.1002/nav.3800020109 91

Kulis, B. and Grauman, K. [2009], Kernelized locality-sensitive hashing, *in* 'Proceedings of the IEEE International Conference on Computer Vision'. 32

Kulis, B., Jain, P. and Grauman, K. [2009], 'Fast similarity search for learned metrics', *IEEE Transactions on Pattern Analysis and Machine Intelligence* **31**(12). DOI: 10.1109/TPAMI.2009.151 xv, 32, 33

Kumar, A. and Sminchisescu, C. [2007], Support kernel machines for object recognition, *in* 'Proceedings of the IEEE International Conference on Computer Vision'. DOI: 10.1109/ICCV.2007.4409065 99

Kumar, M., Torr, P. and Zisserman, A. [2005], OBJ CUT, *in* 'Proceedings of the IEEE Conference on Computer Vision and Pattern Recognition'. DOI: 10.1109/CVPR.2005.249 125, 126

Kumar, N., Berg, A., Belhumeur, P. and Nayar, S. [2009], Attribute and simile classifiers for face verification, *in* 'Proceedings of the IEEE International Conference on Computer Vision'. DOI: 10.1109/ICCV.2009.5459250 129

la Torre Frade, F. D. and Vinyals, O. [2007], Learning kernel expansions for image classification, *in* 'Proceedings of the IEEE Conference on Computer Vision and Pattern Recognition'. DOI: 10.1109/CVPR.2007.383151 99

Labayrade, R. and Aubert, D. [2003], A single framework for vehicle roll, pitch, yaw estimation and obstacles Detection by Stereovision, *in* 'IEEE Intelligent Vehicles Symposium'. DOI: 10.1109/IVS.2003.1212878 122

Ladický, L., Russell, C., Kohli, P. and Torr, P. [2009], Associative hierarchical CRFs for object class image segmentation, *in* 'Proceedings of the IEEE International Conference on Computer Vision'. DOI: 10.1109/ICCV.2009.5459248 125

Ladický, L., Sturgess, P., Alahari, K., Russell, C. and Torr, P. [2010], What, where and how many? Combining object detectors and CRFs, *in* 'Proceedings of the European Conference on Computer Vision'. DOI: 10.1007/978-3-642-15561-1_31 125

Lakoff, G. [1987], *Women, fire, and dangerous things – what categories reveal about the mind*, Univ. of Chicago Press, Chicago. 1

Lampert, C., Blaschko, M. and Hofmann, T. [2008], Beyond sliding windows: Object localization by efficient subwindow search, *in* 'Proceedings of the IEEE Conference on Computer Vision and Pattern Recognition'. DOI: 10.1109/CVPR.2008.4587586 81

Lampert, C., Nickisch, H. and Harmeling, S. [2009], Learning to detect unseen object classes by between-class attribute transfer, *in* 'Proceedings of the IEEE Conference on Computer Vision and Pattern Recognition'. DOI: 10.1109/CVPRW.2009.5206594 127, 129

Lanckriet, G., Cristianini, N., Bartlett, P., Ghaoui, L. E. and Jordan, M. [2004], 'Learning the kernel matrix with semidefinite programming', *Journal of Machine Learning Research* **5**, 27–72. 99

Larlus, D., Verbeek, J. and Jurie, F. [2008], Category level object segmentation by combining bag-of-words models and Markov random fields, *in* 'Proceedings of the IEEE Conference on Computer Vision and Pattern Recognition'. DOI: 10.1007/s11263-009-0245-x 125

Lavrenko, V., Manmatha, R. and Jeon, J. [2003], A model for learning the semantics of pictures, *in* 'Advances in Neural Information Processing Systems'. 128

Lazebnik, S., Schmid, C. and Ponce, J. [2004], Semi-local affine parts for object recognition, *in* 'British Machine Vision Conference'. xvii, 97

Lazebnik, S., Schmid, C. and Ponce, J. [2006], Beyond bags of features: Spatial pyramid matching for recognizing natural scene categories, *in* 'Proceedings of the IEEE Conference on Computer Vision and Pattern Recognition'. DOI: 10.1109/CVPR.2006.68 xvi, 64, 67, 69, 93, 94

LeCun, Y. and Cortes, C., The MNIST database of handwritten digits, http://yann.lecun.com/exdb/mnist/index.html.

Lee, J. J. [2008], LIBPMK: A pyramid match toolkit, MIT Computer Science and Artificial Intelligence Laboratory, http://hdl.handle.net/1721.1/41070.

Lee, Y. J. and Grauman, K. [2008], Foreground focus: Finding meaningful features in unlabeled images, *in* 'British Machine Vision Conference'. 128

Lee, Y. J. and Grauman, K. [2009*a*], 'Foreground focus: Unsupervised learning from partially matching images', *International Journal of Computer Vision*. DOI: 10.1007/s11263-009-0252-y 68, 97, 98, 128

Lee, Y. J. and Grauman, K. [2009*b*], Shape discovery from unlabeled image collections, *in* 'Proceedings of the IEEE Conference on Computer Vision and Pattern Recognition'. DOI: 10.1109/CVPRW.2009.5206698 128

Lee, Y. J. and Grauman, K. [2010*a*], Collect-cut: Segmentation with top-down cues discovered in multi-object images, *in* 'Proceedings of the IEEE Conference on Computer Vision and Pattern Recognition'. DOI: 10.1109/CVPR.2010.5539772 126

Lee, Y. J. and Grauman, K. [2010*b*], Object-graphs for context-aware category discovery, *in* 'Proceedings of the IEEE Conference on Computer Vision and Pattern Recognition (CVPR)', 2010. 123

Lehmann, A., Leibe, B. and Van Gool, L. [2009], PRISM: PRincipled Implicit Shape Model, *in* 'British Machine Vision Conference'. 85

Lehmann, A., Leibe, B. and Van Gool, L. [2010], 'Fast PRISM: Branch and bound Hough transform for object class detection', *International Journal of Computer Vision*. DOI: 10.1007/s11263-010-0342-x 85

Leibe, B., Cornelis, N., Cornelis, K. and Van Gool, L. [2007], Dynamic 3D scene analysis from a moving vehicle, *in* 'Proceedings of the IEEE Conference on Computer Vision and Pattern Recognition'. DOI: 10.1109/CVPR.2007.383146 110, 115, 122, 123, 124

Leibe, B. and Schiele, B. [2003], Analyzing contour and appearance based methods for object categorization, *in* 'Proceedings of the IEEE Conference on Computer Vision and Pattern Recognition'. DOI: 10.1109/CVPR.2003.1211497

Leibe, B., Leonardis, A. and Schiele, B. [2004], Combined object categorization and segmentation with an implicit shape model, *in* 'ECCV'04 Workshop on Statistical Learning in Computer Vision', Prague, Czech Republic. 72, 75, 76

Leibe, B., Leonardis, A. and Schiele, B. [2008], 'Robust object detection with interleaved categorization and segmentation', *International Journal of Computer Vision* **77**(1-3), 259–289. DOI: 10.1007/s11263-007-0095-3 xvii, 72, 75, 76, 83, 84, 100, 110, 111, 112, 113, 125

Leibe, B., Mikolajczyk, K. and Schiele, B. [2006], Segmentation based multi-cue integration for object detection, *in* 'British Machine Vision Conference'. 115

Leibe, B. and Schiele, B. [2003], Interleaved object categorization and segmentation, *in* 'British Machine Vision Conference', Norwich, UK, pp. 759–768. 72, 75, 111, 125

Leibe, B. and Schiele, B. [2004], Scale invariant object categorization using a scale-adaptive mean-shift Search, *in* 'DAGM Annual Pattern Recognition Symposium', Springer LNCS, Vol. 3175, pp. 145–153. DOI: 10.1007/b99676 84

Leibe, B., Schindler, K. and Van Gool, L. [2008], 'Coupled object detection and tracking from static cameras and moving vehicles', *IEEE Transactions on Pattern Analysis and Machine Intelligence* **30**(10), 1683–1698. DOI: 10.1109/TPAMI.2008.170 xv, 51, 122, 123, 124

Leibe, B., Seemann, E. and Schiele, B. [2005], Pedestrian detection in crowded scenes, *in* 'Proceedings of the IEEE Conference on Computer Vision and Pattern Recognition'. DOI: 10.1109/CVPR.2005.272 xvii, 110, 114

Leordeanu, M. and Hebert, M. [2005], A spectral technique for correspondence problems using pairwise constraints, *in* 'Proceedings of the IEEE International Conference on Computer Vision'. DOI: 10.1109/ICCV.2005.20 95

Leordeanu, M., Hebert, M. and Sukthankar, R. [2007], Beyond local appearance: Category recognition from pairwise interactions of simple features, *in* 'Proceedings of the IEEE Conference on Computer Vision and Pattern Recognition'. DOI: 10.1109/CVPR.2007.383091 95

Lepetit, V., Lagger, P. and Fua, P. [2005], Randomized trees for real-time keypoint recognition, *in* 'Proceedings of the IEEE Conference on Computer Vision and Pattern Recognition'. DOI: 10.1109/CVPR.2005.288 31

Leung, T. and Malik, J. [1999], Recognizing surfaces using three-dimensional textons, *in* 'Proceedings of the IEEE International Conference on Computer Vision'. DOI: 10.1109/ICCV.1999.790379 65, 67, 68

Levi, K., Fink, M. and Weiss, Y. [2004], Learning from a small number of examples by exploiting object categories, *in* 'Workshop of Learning in Computer Vision (LCVPR)'. DOI: 10.1109/CVPR.2004.108 127

Li, L.-J., Socher, R. and Fei-Fei, L. [2009], Towards total scene understanding: classification, annotation and segmentation in an automatic framework, *in* 'Proceedings of the IEEE Conference on Computer Vision and Pattern Recognition'. DOI: 10.1109/CVPRW.2009.5206718 128

Li, L., Wang, G. and Fei-Fei, L. [2007], Optimol: Automatic online picture collection via incremental model learning, *in* 'Proceedings of the IEEE Conference on Computer Vision and Pattern Recognition'. DOI: 10.1007/s11263-009-0265-6 128

Lindeberg, T. [1994], 'Scale-space theory: A basic tool for analysing structures at different scales', *Journal of Applied Statistics* **21**(2), 224–270. DOI: 10.1080/757582976 16

Lindeberg, T. [1998], 'Feature detection with automatic scale selection', *International Journal of Computer Vision* **30**(2), 79–116. DOI: 10.1023/A:1008045108935 16, 17, 19

Ling, H. and Soatto, S. [2007], Proximity distribution kernels for geometric context in category recognition, *in* 'Proceedings of the IEEE International Conference on Computer Vision'. DOI: 10.1109/ICCV.2007.4408859 97, 98

Ling, H. and Jacobs, D.W. [2007], Shape classification using the inner-distance, Recognition, *in IEEE Transactions on Pattern Analysis and Machine Intelligence* **29**(2), 286–289. DOI: 10.1109/TPAMI.2007.41 xvi, 90

Liu, D. and Chen, T. [2007], Unsupervised image categorization and object localization using topic models and correspondences between Images, *in* 'Proceedings of the IEEE International Conference on Computer Vision'. DOI: 10.1109/ICCV.2007.4408852 128

Loeff, N. and Farhadi, A. [2008], Scene discovery by matrix factorization, *in* 'Proceedings of the European Conference on Computer Vision'. DOI: 10.1007/978-3-540-88693-8_33 129

Lowe, D. [1999], Object recognition from local scale invariant features, *in* 'Proceedings of the IEEE International Conference on Computer Vision'. xv, xvi, 22, 23, 46, 52, 55, 57, 65

Lowe, D. [2004], 'Distinctive image features from scale-invariant keypoints', *International Journal of Computer Vision* **60**(2), 91–110. DOI: 10.1023/B:VISI.0000029664.99615.94 xv, 5, 9, 10, 17, 18, 21, 22, 23, 25, 28, 34, 51, 55, 57, 65, 91, 107

Lv, F. and Nevatia, R. [2007], Single view human action recognition using key pose matching and Viterbi path searching, *in* 'Proceedings of the IEEE Conference on Computer Vision and Pattern Recognition'. DOI: 10.1109/CVPR.2007.383131 94

Maji, S. and Malik, J. [2009], Object detection using a max-margin hough transform, *in* 'Proceedings of the IEEE Conference on Computer Vision and Pattern Recognition'. DOI: 10.1109/CVPR.2009.5206693 115

Makadia, A., Pavlovic, V. and Kumar, S. [2008], A new baseline for image annotation, *in* 'Proceedings of the European Conference on Computer Vision'. DOI: 10.1007/978-3-540-88690-7_24 129

Malik, J., Belongie, S., Leung, T. and Shi, J. [2001], 'Contour and texture analysis for image segmentation', *International Journal of Computer Vision* **43**(1), 7–27. DOI: 10.1023/A:1011174803800 65, 67

Maron, O. and Ratan, A. [1998], Multiple-instance learning for natural scene classification, *in* 'Proceedings of the International Conference on Machine Learning'. 126

Matas, J. and Chum, O. [2004], 'Randomized RANSAC with T(d,d) test', *Image and Vision Computing* **22**(10), 837–842. DOI: 10.1016/j.imavis.2004.02.009 51

Matas, J. and Chum, O. [2005], Randomized RANSAC with sequential probability ratio test, *in* 'Proceedings of the IEEE International Conference on Computer Vision', pp. 1727–1732. DOI: 10.1109/ICCV.2005.198 51

Matas, J., Chum, O., Martin, U. and Pajdla, T. [2002], Robust wide baseline stereo from maximally stable extremal regions, *in* 'British Machine Vision Conference', pp. 384–393. DOI: 10.1016/j.imavis.2004.02.006 20, 21

Mikolajczyk, C., Schmid, C. and Zisserman, A. [2004], Human detection based on a probabilistic assembly of robust part detectors, *in* 'Proceedings of the European Conference on Computer Vision'. DOI: 10.1007/978-3-540-24670-1_6 101

Mikolajczyk, K., Leibe, B. and Schiele, B. [2006], Multiple object class detection with a generative model, *in* 'Proceedings of the IEEE Conference on Computer Vision and Pattern Recognition'. DOI: 10.1109/CVPR.2006.202 115

Mikolajczyk, K. and Schmid, C. [2001], Indexing based on scale invariant interest points, *in* 'Proceedings of the IEEE International Conference on Computer Vision', pp. 525–531. DOI: 10.1109/ICCV.2001.937561 19, 20

Mikolajczyk, K. and Schmid, C. [2003], A performance evaluation of local descriptors, *in* 'Proceedings of the IEEE Conference on Computer Vision and Pattern Recognition'. DOI: 10.1109/TPAMI.2005.188 20

Mikolajczyk, K. and Schmid, C. [2004], 'Scale & affine invariant interest point detectors', *International Journal of Computer Vision* **60**(1), 63–86. DOI: 10.1023/B:VISI.0000027790.02288.f2 19, 20, 22

Mikolajczyk, K. and Schmid, C. [2005], 'A performance evaluation of local descriptors', *IEEE Transactions on Pattern Analysis and Machine Intelligence* **27**(10), 31–37. DOI: 10.1109/TPAMI.2005.188 11, 23

Mikolajczyk, K., Tuytelaars, T., Schmid, C., Zisserman, A., Matas, J., Schaffalitzky, F., Kadir, T. and Van Gool, L. [2005], 'A comparison of affine region detectors', *International Journal of Computer Vision* **65**(1/2), 43–72. DOI: 10.1007/s11263-005-3848-x 11, 20

Miller, E., Matsakis, N. and Viola, P. [2000], Learning from one example through shared densities on transforms, *in* 'Proceedings of the IEEE Conference on Computer Vision and Pattern Recognition'. DOI: 10.1109/CVPR.2000.855856 127

Mohan, A., Papageorgiou, C. and Poggio, T. [2001], 'Example-based object detection in images by components', *IEEE Transactions on Pattern Analysis and Machine Intelligence* **23**(4), 349–361. DOI: 10.1109/34.917571 83, 101

Monay, F. and Gatica-Perez, D. [2003], On image autoannotation with latent space models, *in* 'ACM Multimedia'. DOI: 10.1145/957013.957070 128

Moosmann, F., Triggs, B. and Jurie, F. [2006], Fast discriminative visual codebooks using randomized clustering forests, *in* 'Advances in Neural Information Processing Systems'. 37, 38

MSR-Cambridge [2005], *Microsoft Research Cambridge Object Recognition Database*, http://research.microsoft.com/en-us/projects/ObjectClassRecognition/. 87, 120

Muja, M. and Lowe, D. [2009], Fast approximate nearest neighbors with automatic algorithm configuration, *in* 'International Conference on Computer Vision Theory and Application'. 28, 29

Murase, H. and Nayar, S. [1995], 'Visual learning and recognition of 3D objects from appearance', *International Journal of Computer Vision* **14**, 5–24. DOI: 10.1007/BF01421486 7

Murilloa, A. and et al. [2007], 'From omnidirectional images to hierarchical localization', *Robotics and Autonomous Systems* **55**(5), 372–382. DOI: 10.1016/j.robot.2006.12.004 94

Murphy, K., Torralba, A., Eaton, D. and Freeman, W. [2006], *Towards category-level object recognition*, LNCS, chapter Object Detection and Localization Using Local and Global Features. 81

Nayar, S., Nene, S. and Murase, H. [1996], Real-time 100 object recognition system, *in* 'Proceedings of ARPA Image Understanding Workshop', San Francisco. DOI: 10.1109/ROBOT.1996.506510 7

Nistér, D. [2003], Preemptive RANSAC for live structure and motion estimation, *in* 'Proceedings of the IEEE International Conference on Computer Vision'. DOI: 10.1007/s00138-005-0006-y 51

Nister, D. and Stewenius, H. [2006], Scalable recognition with a vocabulary tree, *in* 'Proceedings of the IEEE Conference on Computer Vision and Pattern Recognition'. DOI: 10.1109/CVPR.2006.264 37, 59

Nowak, E., Jurie, F. and Triggs, B. [2006], Sampling strategies for bag-of-features image classification, *in* 'Proceedings of the European Conference on Computer Vision'. DOI: 10.1007/11744085_38 67

Obdrzalek, S. and Matas, J. [2005], Sub-linear indexing for large scale object recognition, *in* 'British Machine Vision Conference'. 31

Okuma, K., Taleghani, A., de Freitas, N., Little, J. and Lowe, D. [2004], A boosted particle filter: Multitarget detection and tracking, *in* 'Proceedings of the European Conference on Computer Vision'. DOI: 10.1007/978-3-540-24670-1_3 124

Oliva, A. and Torralba, A. [2001], 'Modeling the shape of the scene: A holistic representation of the spatial envelope', *International Journal of Computer Vision* **42**(3). DOI: 10.1023/A:1011139631724 122

Opelt, A., Pinz, A. and Zisserman, A. [2006*a*], A boundary-fragment-model for object detection, *in* 'Proceedings of the European Conference on Computer Vision', pp. 575–588. DOI: 10.1007/11744047_44 69, 70, 72

Opelt, A., Pinz, A. and Zisserman, A. [2006*b*], Incremental learning of object detectors using a visual alphabet, *in* 'Proceedings of the IEEE Conference on Computer Vision and Pattern Recognition'. DOI: 10.1109/CVPR.2006.153 124

Oxf [2004], 'Oxford interest point webpage', http://www.robots.ox.ac.uk/~vgg/research/affine/detectors.html. 25

Papageorgiou, C. and Poggio, T. [2000], 'A trainable system for object detection', *International Journal of Computer Vision* **38**(1), 15–33. DOI: 10.1023/A:1008162616689 77, 107

Parikh, D., Zitnick, C. L. and Chen, T. [2008], From appearance to context-based recognition: Dense labeling in small images, *in* 'Proceedings of the IEEE Conference on Computer Vision and Pattern Recognition'. DOI: 10.1109/CVPR.2008.4587595 123

Parikh, D., Zitnick, C. L. and Chen, T. [2009], Unsupervised learning of hierarchical spatial structures in images, *in* 'IEEE Conference on Computer Vision and Pattern Recognition'. DOI: 10.1109/CVPRW.2009.5206549 69, 128

Park, D., Ramanan, D. and Fowlkes, C. [2010], Multiresolution models for object detection, *in* 'Proceedings of the European Conference on Computer Vision'. DOI: 10.1007/978-3-642-15561-1_18 118

Parikh, D. and Grauman, K. [2011], Interactively building a discriminative vocabulary of nameable attributes, *in* 'Proceedings of the IEEE Conference in Computer Vision and Patter Recognition. 129

PAS [2007], *http://www.pascal-network.org/challenges/VOC/voc2007/workshop/index.html*. 77

Perronnin, F., Dance, C., Csurka, G. and Bressan, M. [2006], Adapted vocabularies for generic visual categorization, *in* 'Proceedings of the European Conference on Computer Vision'. 38

Philbin, J., Chum, O., Isard, M., Sivic, J. and Zisserman, A. [2007], Object retrieval with large vocabularies and fast spatial matching, *in* 'Proceedings of the IEEE Conference on Computer Vision and Pattern Recognition'. DOI: 10.1109/CVPR.2007.383172 xvi, 5, 37, 57, 58

Philbin, J., Chum, O., Isard, M., Sivic, J. and Zisserman, A. [2008], Lost in quantization: Improving particular object retrieval in large scale image databases, *in* 'Proceedings of the IEEE Conference on Computer Vision and Pattern Recognition'. DOI: 10.1109/CVPR.2008.4587635 37

Philbin, J. and Zisserman, A. [2008], Object mining using a matching graph on very large image collections, *in* 'Proceedings of the Indian Conference on Computer Vision, Graphics and Image Processing'. DOI: 10.1109/ICVGIP.2008.103 128

Proc. IEEE Int'l Workshop "25 Years of RANSAC" in conjunction with CVPR [2006].
URL: *http://cmp.felk.cvut.cz/cvpr06-ransac* 51, 52

Qi, G. J., Hua, X. S. and Zhang, H. J. [2009], Learning semantic distance from community-tagged media collection, *in* 'Multimedia'. DOI: 10.1145/1631272.1631307 129

Quack, T., Ferrari, V., Leibe, B. and Gool, L. V. [2007], Efficient mining of frequent and distinctive feature configurations, *in* 'Proceedings of the IEEE International Conference on Computer Vision'. DOI: 10.1109/ICCV.2007.4408906 xvii, 68, 97, 98, 128

Quack, T., Leibe, B. and Van Gool, L. [2006], World-scale mining of objects and events from community photo collections, *in* 'ACM International Conference on Image and Video Retrieval'. DOI: 10.1145/1386352.1386363 59

Quattoni, A., Collins, M. and Darrell, T. [2007], Learning visual representations using images with captions, *in* 'Proceedings of the IEEE Conference on Computer Vision and Pattern Recognition'. DOI: 10.1109/CVPR.2007.383173 129

Quattoni, A., Collins, M. and Darrell, T. [2008], Transfer learning for image classification with sparse prototype representations, *in* 'Proceedings of the IEEE Conference on Computer Vision and Pattern Recognition'. DOI: 10.1109/CVPR.2008.4587637 127

Rabinovich, A., Vedaldi, A., Galleguillos, C., Wiewiora, E. and Belongie, S. [2007], Objects in context, *in* 'Proceedings of the IEEE International Conference on Computer Vision'. DOI: 10.1109/ICCV.2007.4408986 123

Raginsky, M. and Lazebnik, S. [2009], Locality-sensitive binary codes from shift-invariant kernels, *in* 'Advances in Neural Information Processing Systems'. 33

Raguram, R., Frahm, J.-M. and Pollefeys, M. [2008], A comparative analysis of RANSAC techniques leading to adaptive real-time random sample consensus, *in* 'Proceedings of the European Conference on Computer Vision', pp. 500–513. DOI: 10.1007/978-3-540-88688-4_37 51

Ramanan, D., Forsyth, D. and Zisserman, A. [2007], 'Tracking people by learning their appearance', *IEEE Transactions on Pattern Analysis and Machine Intelligence* **29**(1), 65–81. DOI: 10.1109/TPAMI.2007.250600 76

Razavi, N., Gall, J. and Van Gool, L. [2010], Backprojection revisited: Scalable multi-view object detection and similarity metrics for detections, *in* 'Proceedings of the European Conference on Computer Vision'. DOI: 10.1007/978-3-642-15549-9_45 123

Rohrbach, M., Stark, M., Szarvas, G., Gurevych, I. and Schiele, B. [2010], What helps where – and why? semantic relatedness for knowledge transfer, *in* 'Proceedings of the IEEE Conference on Computer Vision and Pattern Recognition'. DOI: 10.1109/CVPR.2010.5540121 129

Rosch, E., Mervis, C., Gray, W., Johnson, D. and Boyes-Braem, P. [1976], 'Basic objects in natural categories', *Cognitive Psychology* **8**, 382–439. DOI: 10.1016/0010-0285(76)90013-X 1

Rosten, E. and Drummond, T. [2008], Machine learning for high-speed corner detection, *in* 'Proceedings of the European Conference on Computer Vision'. DOI: 10.1007/11744023_34 24

Rother, C., Kolmogorov, V. and Blake, A. [2004], Grabcut: Interactive foreground extraction using iterated graph cuts, *in* 'ACM SIGGRAPH'. DOI: 10.1145/1015706.1015720 125

Rothganger, F., Lazebnik, S., Schmid, C. and Ponce, J. [2003], 3D modeling and recognition using affine-invariant patches and multi-view spatial constraints, *in* 'Proceedings of the IEEE Conference on Computer Vision and Pattern Recognition'. DOI: 10.1109/CVPR.2003.1211480 123

Rowley, H., Baluja, S. and Kanade, T. [1998], 'Neural network-based face detection', *IEEE Transactions on Pattern Analysis and Machine Intelligence* **20**(1), 23–38. DOI: 10.1109/34.655647 77, 80

Russell, B., Efros, A., Sivic, J., Freeman, W. and Zisserman, A. [2006], Using multiple segmentations to discover objects and their extent in image collections, *in* 'Proceedings of the IEEE Conference on Computer Vision and Pattern Recognition'. DOI: 10.1109/CVPR.2006.326 128

Russell, B., Torralba, A., Murphy, K. and Freeman, W. [2008], 'LabelMe: a database and web-based Tool for image annotation', *International Journal of Computer Vision* **77**(1–3), 157–173. DOI: 10.1007/s11263-007-0090-8 87, 100, 111, 120, 121

Russovsky, O. and Ng, A. [2010], A Steiner tree approach for efficient object detection, *in* 'Proceedings of the IEEE Conference on Computer Vision and Pattern Recognition'. DOI: 10.1109/CVPR.2010.5540097 124

Salakhutdinov, R. and Hinton, G. [2007], Semantic hashing, *in* 'ACM SIGIR'. DOI: 10.1016/j.ijar.2008.11.006 33

Sattler, T., Leibe, B. and Kobbelt, L. [2009], SCRAMSAC: Improving RANSAC's efficiency with a spatial consistency filter, *in* 'Proceedings of the IEEE International Conference on Computer Vision'. DOI: 10.1109/ICCV.2009.5459459 51

Savarese, S. and Fei-Fei, L. [2007], 3D generic object categorization, localization, and pose estimation, *in* 'Proceedings of the IEEE International Conference on Computer Vision'. DOI: 10.1109/ICCV.2007.4408987 124

Savarese, S. and Fei-Fei, L. [2008], View synthesis for recognizing unseen poses of object classes, *in* 'Proceedings of the European Conference on Computer Vision'. DOI: 10.1007/978-3-540-88690-7_45 124

Savarese, S., Winn, J. and Criminisi, A. [2006], Discriminative object class models of appearance and shape by correlatons, *in* 'Proceedings of the IEEE Conference on Computer Vision and Pattern Recognition'. DOI: 10.1109/CVPR.2006.102 68, 97

Schaffalitzky, F. and Zisserman, A. [2002], Multi-view matching for unordered image sets, or "How do I organize my holiday snaps?", *in* 'Proceedings of the European Conference on Computer Vision', pp. 414–431. DOI: 10.1007/3-540-47969-4_28 20

Schapire, R., Freund, Y., Bartlett, P. and Lee, W. [1997], Boosting the margin: A new explanation for the effectiveness of voting methods, *in* 'Proceedings of the International Conference on Machine Learning'. 104

Schiele, B. and Crowley, J. [2000], 'Recognition without correspondence using multidimensional receptive field histograms', *International Journal of Computer Vision* **36**(1), 31–52. DOI: 10.1023/A:1008120406972 9

Schmid, C., Mohr, R. and Bauckhage, C. [2000], 'Evaluation of interest point detectors', *International Journal of Computer Vision* **37**(2), 151–172. DOI: 10.1023/A:1008199403446 15

Seemann, E., Leibe, B., Mikolajczyk, K. and Schiele, B. [2005], An evaluation of local shape-based features for pedestrian detection, *in* 'British Machine Vision Conference', Oxford, UK. 115

Seemann, E., Leibe, B., Mikolajczyk, K. and Schiele, B. [2006], Multi-aspect detection of articulated objects, *in* 'Proceedings of the IEEE Conference on Computer Vision and Pattern Recognition'. DOI: 10.1109/CVPR.2006.193 115, 123

Serre, T., Wolf, L. and Poggio, T. [2005], Object recognition with features inspired by visual cortex, *in* 'Proceedings of the IEEE Conference on Computer Vision and Pattern Recognition'. DOI: 10.1109/CVPR.2005.254 64

Shakhnarovich, G. [2005], Learning task-specific similarity, PhD thesis, MIT. 33

Shakhnarovich, G., Darrell, T. and Indyk, P., eds [2006], *Nearest-neighbor methods in learning and vision: Theory and practice*, MIT Press. DOI: 10.1109/ICCV.2003.1238424 32

Shakhnarovich, G., Viola, P. and Darrell, T. [2003], Fast pose estimation with parameter-sensitive hashing, *in* 'Proceedings of the IEEE International Conference on Computer Vision'. 33, 99

Shi, J. and Malik, J. [2000], 'Normalized cuts and image segmentation', *IEEE Transactions on Pattern Analysis and Machine Intelligence* **22**(8), 888–905. DOI: 10.1109/34.868688 125

Shotton, J., Winn, J., Rother, C. and Criminisi, A. [2006], TextonBoost: joint appearance, shape and context modeling for multi-class object recognition and segmentation, *in* 'Proceedings of the European Conference on Computer Vision'. 68, 82, 123

Siddiquie, B. and Gupta, A. [2010], Beyond active noun tagging: Modeling contextual interactions for multi-class active learning, *in* 'Proceedings of the IEEE Conference on Computer Vision and Pattern Recognition'. DOI: 10.1109/CVPR.2010.5540044 127

Silpa-Anan, C. and Hartley, R. [2008], Optimised kd-trees for fast image descriptor matching, *in* 'Proceedings of the IEEE Conference on Computer Vision and Pattern Recognition'. DOI: 10.1109/CVPR.2008.4587638 28, 29

Singhal, A., Luo, J. and Zhu, W. [2003], Probabilistic spatial context models for scene content understanding, *in* 'Proceedings of the IEEE Conference on Computer Vision and Pattern Recognition'. DOI: 10.1109/CVPR.2003.1211359 82, 123

Sivic, J., Russell, B., Efros, A., Zisserman, A. and Freeman, W. [2005], Discovering object categories in image collections, *in* 'Proceedings of the IEEE International Conference on Computer Vision'. 128

Sivic, J. and Zisserman, A. [2003], Video Google: A text retrieval approach to object matching in videos, *in* 'Proceedings of the IEEE International Conference on Computer Vision', Nice. DOI: 10.1109/ICCV.2003.1238663 xv, 36, 37

Sivic, J. and Zisserman, A. [2004], Video data mining using configurations of viewpoint ivariant regions, *in* 'Proceedings of the IEEE Conference on Computer Vision and Pattern Recognition', Washington, D.C. DOI: 10.1109/CVPR.2004.1315071 xvii, 97

Stark, M., Goesele, M. and Schiele, B. [2009], A shape-based object class model for knowledge transfer, *in* 'Proceedings of the IEEE International Conference on Computer Vision'. DOI: 10.1109/ICCV.2009.5459231 127

Strecha, C., A. M. Bronstein, M. M. B. and Fua, P. [2010], LDAHash: Improved matching with smaller descriptors, *in* 'EPFL-REPORT-152487'. 33

Su, H., Sun, M., Fei-Fei, L. and Savarese, S. [2009], Learning a dense multi-view representation for detection, viewpoint classification and synthesis of object categories, *in* 'Proceedings of the IEEE International Conference on Computer Vision'. DOI: 10.1109/ICCV.2009.5459168 124

Sudderth, E., Torralba, A., Freeman, W. and Willsky, A. [2005], Learning hierarchical models of scenes, objects, and parts, *in* 'Proceedings of the IEEE International Conference on Computer Vision'. DOI: 10.1109/ICCV.2005.137 69

SUR [2006], 'SURF features website', http://www.vision.ee.ethz.ch/~surf. 25

Swain, M. and Ballard, D. [1991], 'Color indexing', *International Journal of Computer Vision* **7**(1), 11–32. DOI: 10.1007/BF00130487 9

Thayananthan, A., Stenger, B., Torr, P. H. S. and Cipolla, R. [2003], Shape context and chamfer matching in cluttered scenes, *in* 'Proceedings of the IEEE Conference on Computer Vision and Pattern Recognition'. DOI: 10.1109/CVPR.2003.1211346 95

Thomas, A., Ferrari, V., Leibe, B., Tuytelaars, T., Schiele, B. and Van Gool, L. [2006], Towards multi-view object class detection, *in* 'Proceedings of the IEEE Conference on Computer Vision and Pattern Recognition'. DOI: 10.1109/CVPR.2006.311 115, 123

Thomas, A., Ferrari, V., Leibe, B., Tuytelaars, T. and Van Gool, L. [2007], Depth-from-recognition: Inferring meta-data through cognitive feedback, *in* 'ICCV Workshop on 3D Representations for Recognition', Rio de Janeiro, Brazil. DOI: 10.1109/ICCV.2007.4408831 112, 113

Thomas, A., Ferrari, V., Leibe, B., Tuytelaars, T. and Van Gool, L. [2009a], 'Shape-from-recognition: Recognition enables meta-data transfer', *Computer Vision and Image Understanding*. (to appear). DOI: 10.1016/j.cviu.2009.03.010 xvii, 112, 113, 125

Thomas, A., Ferrari, V., Leibe, B., Tuytelaars, T. and Van Gool, L. [2009b], 'Using multi-view recognition to guide a robot's attention', *International Journal of Robotics Research* **28**(8), 976–998. 123

Tieu, K. and Viola, P. [2000], Boosting image retrieval, *in* 'Proceedings of the IEEE Conference on Computer Vision and Pattern Recognition'. 99

Todorovic, S. and Ahuja, N. [2006], Extracting subimages of an unknown category from a set of images, *in* 'Proceedings of the IEEE Conference on Computer Vision and Pattern Recognition'. DOI: 10.1109/CVPR.2006.116 126

Torr, P. and Zisserman, A. [2000], 'MLESAC: a new robust estimator with application to estimating image geometry', *Computer Vision and Image Understanding* **78**(1), 138–156. DOI: 10.1006/cviu.1999.0832 51

Torralba, A. [2003], 'Contextual priming for object detection', *International Journal of Computer Vision* **53**(2), 169–191. DOI: 10.1023/A:1023052124951 xvi, 64, 65, 81, 82, 122

Torralba, A., Fergus, R. and Weiss, Y. [2008], Small codes and large image databases for recognition, *in* 'Proceedings of the IEEE Conference on Computer Vision and Pattern Recognition'. DOI: 10.1109/CVPR.2008.4587633 33

Torralba, A., Murphy, K. and Freeman, W. [2004], Sharing features: Efficient boosting procedures for multiclass object detection, *in* 'Proceedings of the IEEE Conference on Computer Vision and Pattern Recognition'. DOI: 10.1109/CVPR.2004.1315241 70, 124

Torralba, A., Murphy, K., Freeman, W. and Rubin, M. [2003], Context-based vision system for place and object recognition, *in* 'Proceedings of the IEEE International Conference on Computer Vision'. DOI: 10.1109/ICCV.2003.1238354 122

Torralba, A., Oliva, A., Castelhano, M. and Henderson, J. [2006], 'Contextual guidance of attention in natural scenes: The role of global features on object search', *Psychological Review*. 122

Triggs, B. [2004], Detecting keypoints with stable position, orientation and scale under illumination changes, *in* 'Proceedings of the European Conference on Computer Vision'. DOI: 10.1007/b97873 13

Tu, Z., Chen, X., Yuille, A. and Zhu, S.-C. [2003], Image parsing: Unifying segmentation, detection, and recognition, *in* 'Proceedings of the IEEE International Conference on Computer Vision'. DOI: 10.1109/ICCV.2003.1238309 125

Turk, M. and Pentland, A. [1992], Face recognition using eigenfaces, *in* 'Proceedings of the IEEE Conference on Computer Vision and Pattern Recognition', pp. 586–590. DOI: 10.1109/CVPR.1991.139758 xvi, 7, 64

Tuytelaars, T. and Mikolajczyk, K. [2007], 'Local invariant feature detectors: A survey', *Foundations and Trends in Computer Graphics and Vision* **3**(3), 177–280. DOI: 10.1561/0600000017 xv, 11, 18, 19, 21, 22

Tuytelaars, T. and Van Gool, L. [2000], Wide baseline stereo matching based on local, affinely invariant regions, *in* 'British Machine Vision Conference', Bristol, UK, pp. 412–422. 20

Tuytelaars, T. and Van Gool, L. [2004], 'Matching widely separated views based on affinely invariant neighbourhoods', *International Journal of Computer Vision* **59**(1), 61–85. DOI: 10.1023/B:VISI.0000020671.28016.e8 xv, 20, 21, 55, 56

Uhlmann, J. [1991], 'Satisfying general proximity / similarity queries with metric trees', *Information Processing Letters* **40**, 175–179. DOI: 10.1016/0020-0190(91)90074-R 29

Varma, M. and Ray, D. [2007], Learning the discriminative power-invariance trade-off, *in* 'Proceedings of the IEEE International Conference on Computer Vision'. DOI: 10.1109/ICCV.2007.4408875 99

Varma, M. and Zisserman, A. [2002], Classifying images of materials: Achieving viewpoint and illumination independence, *in* 'Proceedings of the European Conference on Computer Vision'. DOI: 10.1007/3-540-47977-5_17 68

Veltkamp, R. and Hagedoorn, M. [1999], State-of-the-art in shape matching, *in* 'Tech Report UU-CS-1999-27', Utrecht University. 90

Vijayanarasimhan, S. and Grauman, K. [2008*a*], Keywords to visual categories: Multiple-instance learning for weakly supervised object categorization, *in* 'Proceedings of the IEEE Conference on Computer Vision and Pattern Recognition'. 126

Vijayanarasimhan, S. and Grauman, K. [2008*b*], Multi-level active prediction of useful image annotations for recognition, *in* 'Advances in Neural Information Processing Systems'. 127

Vijayanarasimhan, S. and Grauman, K. [2009], What's it going to cost you?: Predicting effort vs. informativeness for multi-label image annotations., *in* 'Proceedings of the IEEE Conference on Computer Vision and Pattern Recognition'. DOI: 10.1109/CVPRW.2009.5206705 127

Vijayanarasimhan, S. and Grauman, K. [2011], Cost-sensitive active visual category learning, *International Journal of Computer Vision* **91**(1), 24. DOI: 10.1007/s11263-010-0372-4 xvi, 89

Viola, P. and Jones, M. [2001], Rapid object detection using a boosted cascade of simple features, *in* 'Proceedings of the IEEE Conference on Computer Vision and Pattern Recognition'. DOI: 10.1109/CVPR.2001.990517 xv, xvi, 5, 81, 82, 89, 103, 105

Viola, P. and Jones, M. [2004], 'Robust real-time face detection', *International Journal of Computer Vision* **57**(2), 137–154. DOI: 10.1023/B:VISI.0000013087.49260.fb xvi, 24, 65, 67, 70, 76, 77, 80, 81

Viola, P., Platt, J. and Zhang, C. [2005], Multiple instance boosting for object detection, *in* 'Advances in Neural Information Processing Systems'. 126

Wang, J., Kumar, S. and Chang, S.-F. [2010], Semi-supervised hashing for scalable image retrieval, *in* 'Proceedings of the IEEE Conference on Computer Vision and Pattern Recognition'. DOI: 10.1109/CVPR.2010.5539994 33

Wang, J., Markert, K. and Everingham, M. [2009], Learning models for object recognition from natural language descriptions, *in* 'British Machine Vision Conference'. 129

Wang, J., Yang, J., Yu, K., Lv, F., Huang, T. and Gong, Y. [2010], Locality-constrained linear coding for image classification, *in* 'Proceedings of the IEEE Conference on Computer Vision and Pattern Recognition'. DOI: 10.1109/CVPR.2010.5540018 109

Wang, Y. and Mori, G. [2010], A discriminative latent model of object classes and attributes, *in* 'Proceedings of the European Conference on Computer Vision'. DOI: 10.1007/978-3-642-15555-0_12 129

Weber, M., Welling, M. and Perona, P. [2000*a*], Towards automatic discovery of object categories, *in* 'Proceedings of the IEEE Conference on Computer Vision and Pattern Recognition'. DOI: 10.1109/CVPR.2000.854754 73

Weber, M., Welling, M. and Perona, P. [2000*b*], Unsupervised learning of object models for recognition, *in* 'Proceedings of the European Conference on Computer Vision'. DOI: 10.1007/3-540-45054-8_2 73, 91, 126

Winn, J., Criminisi, A. and Minka, T. [2005], Object categorization by learned universal visual dictionary, *in* 'Proceedings of the IEEE International Conference on Computer Vision'. DOI: 10.1109/ICCV.2005.171 38

Winn, J. and Jojic, N. [2005], LOCUS: Learning Object Classes with Unsupervised Segmentation, *in* 'Proceedings of the IEEE International Conference on Computer Vision'. DOI: 10.1109/ICCV.2005.148 126

Witkin, A. [1983], Scale-space filtering, *in* 'Proceedings IJCAI', Karlsruhe, Germany, pp. 1019–1022. 16

Wu, B. and Nevatia, R. [2007], 'Detection and tracking of multiple, partially occluded humans by Bayesian Combination of edgelet part detectors', *International Journal of Computer Vision* **75**(2), 247–266. DOI: 10.1007/s11263-006-0027-7 124

Xu, D., Cham, T., Yan, S. and Chang, S.-F. [2008], Near duplicate image identification with spatially aligned pyramid matching, *in* 'Proceedings of the IEEE Conference on Computer Vision and Pattern Recognition'. DOI: 10.1109/CVPR.2008.4587720 94

Yakhnenko, O. and Honavar, V. [2009], Multiple label prediction for image annotation with multiple kernel correlation models, *in* 'Workshop on Visual Context Learning, in conjunction with CVPR'. DOI: 10.1109/CVPR.2009.5204274 129

Yang, C. and Lozano-Perez, T. [2000], Image database retrieval with multiple-instance learning techniques, *in* 'ICDE'. DOI: 10.1109/ICDE.2000.839416 126

Yang, J., Yu, K., Gong, Y. and Huang, T. [2009], Linear spatial pyramid matching sparse coding for image classification, *in* 'Proceedings of the IEEE Conference on Computer Vision and Pattern Recognition'. 109

Yang, L. [2006], Distance metric learning: A comprehensive survey, Technical report, Michigan State Univ. 98

Yang, L., Jin, R., Sukthankar, R. and Jurie, F. [2008], Discriminative visual codebook generation with classifier training for object category recognition, *in* 'Proceedings of the IEEE Conference on Computer Vision and Pattern Recognition'. 38

Yeh, T., Lee, J. and Darrell, T. [2007], Adaptive vocabulary forests for dynamic indexing and category learning, *in* 'Proceedings of the IEEE International Conference on Computer Vision'. DOI: 10.1109/ICCV.2007.4409053 37

Yu, S., Gross, R. and Shi, J. [2002], Concurrent object recognition and segmentation by graph partitioning, *in* 'Advances in Neural Information Processing Systems'. 126

Yu, S. and Shi, J. [2003], Object-specific figure-ground segregation, *in* 'Proceedings of the IEEE Conference on Computer Vision and Pattern Recognition'. DOI: 10.1109/CVPR.2003.10006 124

Yuan, J., Wu, Y. and Yang, M. [2007], Discovery of collocation patterns: from visualwords to visual phrases, *in* 'Proceedings of the IEEE Conference on Computer Vision and Pattern Recognition'. DOI: 10.1109/CVPR.2007.383222 97

Yuille, A., Cohen, D. and Hallinan, P. [1989], Feature extraction from faces using deformable templates, *in* 'Proceedings of the IEEE Conference on Computer Vision and Pattern Recognition'. DOI: 10.1109/CVPR.1989.37836 89

Yuille, A., Hallinan, P., and Cohen, D. [1992], Feature extraction from faces using deformable templates, *International Journal of Computer Vision*, (2), 99–111. DOI: 10.1007/BF00127169 xvi, 89

Zha, Z.-J., Hua, X.-S., Mei, T., Wang, J., Qi, G.-J. and Wang, Z. [2008], Joint multi-label multi-instance learning for image classification, *in* 'Proceedings of the IEEE Conference on Computer Vision and Pattern Recognition'. DOI: 10.1109/CVPR.2008.4587384 127

Zhang, M. L. and Zhou, Z. H. [2007], Multi-label learning by instance differentiation, *in* 'AAAI'. 127

Zhang, Q. and Goldman, S. [2002], EM-DD: An improved multiple-instance learning technique, *in* 'Advances in Neural Information Processing Systems'. 126

Zhou, Z. H. and Zhang, M. L. [2006], Multi-instance multi-label learning with application to scene classification, *in* 'Advances in Neural Information Processing Systems'. 127

Authors' Biographies

KRISTEN GRAUMAN

Kristen Grauman is the Clare Boothe Luce Assistant Professor in the Department of Computer Science at the University of Texas at Austin. Her research focuses on object recognition and visual search. Before joining UT-Austin in 2007, she received her Ph.D. in Computer Science from the Massachusetts Institute of Technology (2006), and a B.A. in Computer Science from Boston College (2001). Grauman has published over 40 articles in peer-reviewed journals and conferences, and work with her colleagues on large-scale visual search received the Best Student Paper Award at the IEEE Conference on Computer Vision and Pattern Recognition (CVPR) in 2008. She is a Microsoft Research New Faculty Fellow, a recipient of an NSF CAREER award and the Howes Scholar Award in Computational Science, and was named one of "AI's Ten to Watch" in IEEE Intelligent Systems in 2011. She serves regularly on the program committees for the major computer vision conferences and is a member of the editorial board for the International Journal of Computer Vision.

BASTIAN LEIBE

Bastian Leibe is an Assistant Professor at RWTH Aachen University. He holds an M.Sc. degree from Georgia Institute of Technology (1999), a Diploma degree from the University of Stuttgart (2001), and a Ph.D. from ETH Zurich (2004), all three in Computer Science. After completing his dissertation on visual object categorization at ETH Zurich, he worked as a postdoctoral research associate at TU Darmstadt and at ETH Zurich. His main research interest are in object categorization, detection, segmentation, and tracking, as well as in large-scale image retrieval and visual search. Bastian Leibe has published over 60 articles in peer-reviewed journals and conferences. Over the years, he received several awards for his research work, including the Virtual Reality Best Paper Award in 2000, the ETH Medal and the DAGM Main Prize in 2004, the CVPR Best Paper Award in 2007, the DAGM Olympus Prize in 2008, and the ICRA Best Vision Paper Award in 2009. He serves regularly on the program committee of the major computer vision conferences and is on the editorial board of the Image and Vision Computing journal.

Printed in the United States
by Baker & Taylor Publisher Services